Action Learning and Action Research Journal

Vol 31 No 2 December 2025

Action Learning, Action Research Association Ltd (and its predecessors) has published the ALAR Journal since 1996.

Managing Editor: Dr Yedida Bessemer

Global Strategic Publications Editorial Board:
Prof John Andersen, Roskilde University, Denmark
Dr Azril Bacal, University of Uppsala, Sweden
Dr Christina Marouli, The American College of Greece, Greece
Prof Imelda Smidt, North-West University, South Africa
Adjunct Professor Yedida Bessemer, University of Charleston, West Virginia, USA

Editorial inquiries:
The Editor, *ALAR Journal*
Action Learning, Action Research Association Ltd
PO Box 164, Crows Nest, NSW 1585 Australia

editor@alarassociation.org

ISSN 1326-964X (Print)   ISSN 2206-611X (Online)

The Action Learning and Action Research Journal is listed in:

- Australian Research Council – *Excellence in Research for Australia (ERA) 2018 Journal List*
- Australian Business Deans Council - *2019 ABDC Journal Quality List*
- Norwegian Directorate for Higher Education and Skills - *Norwegian Register for Scientific Journals, Series and Publishers*

Editorial Advisory Board

Prof John Andersen	Denmark
Dr Rajiv George Aricat	India
Dr Noa Avriel-Avni	Israel
Dr Azril Bacal Roji	Sweden / Chile
Dr Yedida Bessemer	USA / Israel
Colin Bradley	Australia
Dr Daniela Cialfi	Italy
Dr Daniel Cisneiros	Brazil
Dr Philip Crane	Australia
Dr Bob Dick	Australia
Dr Kathryn Epstein	USA
Dr Terrance Fernsler	USA
Dr Susan Goff	Australia
Assoc. Prof. Marina Harvey	Australia
Dr Geof Hill	Australia
Ms Jane Holloway	Australia
Dr Marie Huxtable	UK
Dr Edward Hyatt	USA
Prof. Vasudha Kamat	India
Dr Hanen Khanchel	Tunisia
Dr Elyssabeth Leigh	Australia
Prof Christina Marouli	Greece
Dr John Molineux	Australia
Dr Sumesh Nair	Australia
Ms Margaret O'Connell	Australia
Dr Lizana Oberholzer	UK
Prof Akihiro Ogawa	Australia
Dr Chin Lye Ooi	Malaysia
Dr Paul Pettigrew	UK
Dr Eileen Piggot-Irvine	New Zealand
Ms Riripeti Reedy	New Zealand
Prof Shankar Sankaran	Australia
Dr Natalie Smith	Australia

Dr Steve Smith	Australia
Em Prof Nadarajah Sriskandarajah	Sweden
Prof Emmanuel Tetteh	USA
Assoc Prof Maria Tibon	Philippines
Dr Abbie Victoria Trott	Australia
Dr Anienie Veldsman	New Zealand
Robert Wanyama	Kenya
Assoc Prof Hilary Whitehouse	Australia
Prof Lesley Wood	South Africa

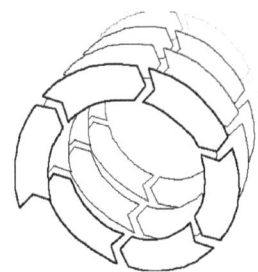

ALAR Journal

Volume 31 No 2
December 2025

ISSN 1326-964X (Print)
ISSN 2206-611X (Online)

CONTENTS

Editorial	7
Yedida Bessemer	
A contextually adaptive data collaborative using action design research	12
Emmanuel Candido Soriente Santos, Renato Andrin Wilano, Jonathan Moss & Benjamin Wilson	
Action research and transformative learning in cross-sector healthcare collaboration	47
Heidi Lene Myglegård Andersen, John Andersen & Ditte Høgsgaard	
It takes a village: Developing action researchers through a transdisciplinary peer-learning collaborative	73
Keith Heggart, Susanne Pratt, Shankar Sankaran and Pernille H. Christensen	

Enhancing EFL teacher participation in an asynchronous online forum: Integrating the Delphi technique with emancipatory participatory action research 112

Stuart Warrington

Book review - The politics of action research: A story telling inquiry 144

Yedida Bessemer

Membership information and article submissions 149

© 2025. Action Learning, Action Research Association Ltd and the author(s) jointly hold the copyright of *ALAR Journal* articles.

Editorial

Greetings

Action research continues to expand through its fundamental commitments to democratic participation, relational learning, and transformative change, but faces persistent challenges from institutional structures that privilege conventional research paradigms. The works included in this issue offer a compelling response, demonstrating how action research thrives when practitioners embrace its inherently political nature, cultivate communities of learning, and design for socio-technical complexity.

Together, these four articles and the featured book review shed light on a central theme: action research is not merely a methodology but a political and relational practice that demands courage, creativity, and collective effort to navigate institutional constraints while generating meaningful change. These works show how action researchers are reimagining what participatory inquiry can achieve in complex, digitally mediated spaces through data collaboratives, healthcare coordination, doctoral education, and online teacher communities.

The first article, A contextually adaptive data collaborative using action design research (Santos et al.), examines action design research in public sector data collaboratives, contrasting Australia's failed Robodebt scheme with the successful South Australia (SA) BLADE initiative. The authors demonstrate that technical feasibility alone cannot guarantee success. Instead, sustained socio-technical co-evolution through iterative collaboration proves essential. This multi-cycle ADR approach shows how trust, learning, and user participation enable system adoption over time, reinforcing a crucial lesson: meaningful data innovation requires human-centered, iterative inquiry, especially as AI and machine learning increasingly shape policy environments.

The second article, Action research and transformative learning in cross-sector healthcare collaboration (Andersen, Andersen, & Høgsgaard), demonstrates how participatory action research combined with digital facilitation can transform healthcare delivery. The Virtual 4-Meetings project brings together patients, relatives, and diverse healthcare professionals for shared care planning, addressing fragmented systems through the Circle-Care model. The researchers' insider roles enabled trust-building and co-creation across professional boundaries, while revealing a sobering reality: action research innovations rarely sustain without leadership support and institutional anchoring, a pattern echoed throughout this collection.

The third article, It takes a village: Developing action researchers through a transdisciplinary peer learning collaborative (Heggart et al.), describes the PAR Collaborative, which offers a compelling alternative to traditional supervisory models that marginalize participatory research methods. Through distributed mentorship, cross-disciplinary dialogue, and scaffolded learning, this year-long program demonstrates that "it takes a village" to develop action researchers. Participants reported increased confidence and reduced isolation, yet the program's reliance on facilitator goodwill rather than institutional infrastructure reveals the precarious position of such initiatives within systems driven by performance metrics and risk aversion.

The fourth article, Enhancing EFL teacher participation in an asynchronous online forum: Integrating the Delphi technique with emancipatory participatory action research (Warrington), addresses unequal participation in online teacher forums by integrating the Delphi technique within Emancipatory Participatory Action Research. This methodological innovation uses anonymity and iterative feedback to mitigate power imbalances among Japanese EFL teachers, demonstrating how emancipatory aims can be pursued in digitally mediated contexts. The study's honest portrayal of emancipatory change as gradual and relational strengthens its credibility, offering actionable insights for culturally sensitive professional learning communities.

The fifth piece is a book review of Hill and Rixon's book, *The Politics of Action Research: A Storytelling Inquiry* (2024). This book provides essential critical framing for understanding these empirical studies. Drawing on narratives from practitioners across continents, the authors argue that action research is inherently political, embedded in power relations that shape who can research, what counts as knowledge, and whose voices are legitimized. Bob Dick's foreword distinguishes between viewing action research as "apolitical" versus recognizing it as embodying a different politics: one characterized by egalitarian participation rather than hierarchical control.

The book's storytelling methodology reveals experiences of empowerment and disempowerment, gatekeeping, and resistance, offering three key contributions. First, methodological transparency demonstrates that narrative data can generate rigorous insights. Second, navigational strategies for "working in the swamp" of messy practice contexts model creative resistance to constraining norms. Third, centering lived experience legitimizes practitioner expertise and honors multiple knowledge forms, embodying action research's commitment to "extended epistemologies."

Readers can see how the featured articles form insightful links. The data collaborative navigates institutional politics around governance. The healthcare project confronts professional hierarchies. The doctoral program challenges university metrics and traditions. The online forum addresses cultural power dynamics. Each exemplifies the political work of action research by claiming legitimacy for participatory knowledge and creating spaces for collaborative inquiry within constraining structures.

However, the book's theoretical density may ironically limit accessibility for practitioners by mirroring the institutional barriers it critiques. Additionally, while acknowledging diverse traditions, deeper engagement with decolonial and Indigenous methodologies would strengthen the analysis, particularly given the PAR Collaborative's efforts to incorporate Indigenous approaches.

As we look to the future, collectively, these works offer critical insights for contemporary practice. First, embrace methodological hybridity. The integration of action design research with socio-technical theory, PAR with digital facilitation, and EPAR with Delphi techniques demonstrates that action research is strengthened through creative synthesis rather than methodological purity. Second, design for sustained engagement. Each study reveals that lasting change requires extended timeframes for trust-building and adaptation through multi-cycle implementations, year-long programs, and iterative forum development. Action researchers must advocate for institutional structures that support sustained engagement rather than rapid pilots.

A third insight is the center relational capacity. Technical solutions and methodological rigor matter, but without attention to relationships, power dynamics, and collective learning, interventions falter. Relational capacity is foundational, not supplementary. The fourth insight is to make political dynamics explicit. By naming gatekeeping, addressing power imbalances, and redistributing voice, action researchers can work more effectively toward transformative change. Lastly, leverage digital affordances thoughtfully. Digital technologies enable new forms of participation and collaboration, yet without careful attention to design and cultural context, they may reproduce rather than challenge existing inequities.

All in all, these featured works demonstrate that action research continues evolving, adapting to digital transformation and data-intensive environments while maintaining core commitments to democratic participation and transformative change. Yet this evolution requires courage to challenge institutional norms, claim legitimacy for participatory knowledge, center relationships alongside technical expertise, and persist despite gatekeeping and resource constraints.

For practitioners, these works offer validation: your experiences navigating institutional politics, building trust, and sustaining collaboration are not individual failures but structural features of the terrain. The methodological innovations documented here provide concrete strategies for working effectively within these constraints.

For supervisors and program directors: How can you create conditions that support rather than hinder action research? This may mean advocating for revised ethics protocols, recognizing diverse scholarly contributions, and allocating resources for sustained engagement.

For the broader field: Are we willing to confront the political dimensions Hill and Rixon illuminate? Can we build the "villages" necessary while transforming university systems that marginalize participatory approaches? Will we engage seriously with decolonial critiques and Indigenous methodologies?

As you engage with the full articles, read not just for methodological techniques but also for the political and relational work they represent. Consider how their insights might inform your practice. Reflect on the communities that sustain your work and the structures that enable or constrain it.

Action research has never promised easy answers. What it offers is a framework for collective learning, ethical commitment, and persistent efforts toward more just futures. These works demonstrate that this framework remains essential as we navigate the complexity of contemporary research. Let us move forward with both realism and hope by acknowledging constraints while remaining committed to transformative possibilities that emerge when we research *with* rather than *on*, privilege relationships alongside technical excellence, and claim space for voices too often silenced.

To Lifelong Learning,
Dr. Yedida Bessemer

A contextually adaptive data collaborative using action design research

Emmanuel Candido Soriente Santos[1], Renato Andrin Villano[2], Jonathan Moss[2] & Benjamin Wilson[1]

Abstract

This paper explores attributes of a successful data innovation project, which challenges the conventional focus on technical feasibility by highlighting the paramount importance of data intelligibility, relevance, and utility in a socio-technical system. The South Australian (SA) data linkage project in the Business Longitudinal Analytical Data Environment (BLADE) is the first of its kind in Australia where a State government has linked business-related administrative data with Commonwealth data. The data linkage has created a new data asset that opens new opportunities for users in producing policy insights. This study sought to show the impact of this innovation using action design research. An important lesson learnt was that collaboration with partners and stakeholders, particularly the end-users is key to fostering a socio-technical system of co-evolution that mutually informs the other process to avoid adverse unintended consequences. This is even more important with the emergence of artificial intelligence and machine learning.

Key words: Action design research, artificial intelligence, data collaborative, data linkage, economic research, machine learning

---

1   Department of the Premier and Cabinet, Government of South Australia
2   University of New England Business School, Armidale, NSW

What is known about the topic?
The importance of aligning technical artefacts with social systems is a foundational concept in socio-technical systems (STS) theory. Data collaboratives are a new and emerging model for handling complex data analytical projects involving multiple partners and sources of data. Despite the availability of data and enabling legal and technical frameworks for undertaking data linkage projects, very few data collaboratives make it past the pilot stage.
What does this paper add?
The paper provides a case study of a successful, first-of-its kind data collaborative involving multiple partners from different jurisdictions in Australia. It highlights the importance of deep collaboration in the governance of the initiative to ensure survivability and replicability, and the distinct advantage of using an action design research approach for facilitating this.
Who will benefit from its content?
Practitioners of data science, policy and project leaders who commission and design governance frameworks for data initiatives, Artificial Intelligence and data transformation strategists and managers.
What is the relevance to AL and AR scholars and practitioners?
This paper demonstrates how action design research (a fusion of action research and design science research) is well-suited for promoting the co-evolution of social and technical systems to ensure the successful iteration and durability of data collaboratives beyond the pilot stage.

Received   April 2025   Reviewed July 2025   Published December 2025

Introduction

The increasing reliance on data analytics and artificial intelligence (AI) to inform public policy presents a dual potential: unprecedented efficiency and catastrophic failure. The latter was starkly illustrated by Australia's 'Robodebt' scheme, where an automated algorithm, designed by external consultants to detect welfare fraud, resulted in immense human and economic cost (Commonwealth of Australia 2023, xxix). The Royal Commission into the scheme concluded that its failure was fundamentally rooted in a lack of stakeholder consultation and an unwillingness to heed feedback from front-line staff, a classic case of a technical system being imposed upon, rather than co-developed with, its social context.

The Robodebt case highlights a critical tension in the Information Systems (IS) and public administration literature. While the importance of aligning technical artefacts with social systems is a foundational concept in socio-technical systems (STS) theory, much of the research focuses on the outcomes of this alignment or misalignment. There remains a significant gap in empirical studies that meticulously unpack the *process* by which this alignment is achieved during the formative stages of complex, multi-agency data innovation projects. It is often in the iterative, and frequently contested, interactions between developers, data custodians, and end-users that the ultimate success or failure of an innovation is determined.

This paper addresses this gap by presenting a case study of a successful data innovation: the South Australian Business Longitudinal Analytical Data Environment (SA BLADE) project. This project represents the first instance in Australia of a State government linking its administrative data with the Commonwealth's BLADE platform to create a new data asset for sub-national policy analysis. Using an Action Design Research (ADR) methodology, this study analyses the project's first three implementation cycles (FY2017/18– FY2021/22) to answer the following research question:

> How does a collaborative, multi-cycle Action Design Research (ADR) approach facilitate the co-evolution of social and technical systems to ensure the successful implementation and adoption of a sub-national linked data asset?

This paper argues that the success of SA BLADE was contingent not on its technical feasibility alone, but on the ADR process that fostered a continuous, adaptive feedback loop between policy end-users (the social system) and the data asset's development (the technical system). This iterative process built inter-agency trust, revealed unanticipated use-cases that demonstrated immediate value and ensured the resulting IS artefact was intelligible, relevant and ultimately fit-for-purpose. In contrast to the top-down imposition of Robodebt, SA BLADE demonstrates a model of co-evolution where human participation was not a barrier to be managed, but the core mechanism for innovation.

The structure of the paper will be as follows: Part 1 presents the contextual challenges faced by the project proponent, necessitating data innovation. Part 2 provides a brief review of the literature. Part 3 describes the data innovation, process, stakeholders, and intended outcomes of the project. This is followed by the framing and results of the research into what constitutes success (Part 4) and a discussion of the wider implications of the project and research study to both theory and practice (Part 5). The conclusion provides some closing remarks and reflections.

Part 1: Contextual challenges

Governments at all levels recognise the potential to utilise public data to address complex policy questions, and the South Australian Government (SAG) wished to better understand the trends and drivers of growth within the State to provide an evidence base for its socioeconomic plans. The Department of the Premier and Cabinet (DPC) is the lead agency in developing policy and delivering programs to realise the government's vision for South Australia. The DPC sought to develop an overall data and analytical framework and capability across the SAG to generate better insights that could support the design, monitoring and appraisal of its programs. It was essential that this framework be reliable, responsive and fit for multiple purposes over a range of policy domains related to the SAG's priorities as they evolved.

There are existing published data that are relevant, but all have their limitations in terms of reliability, coverage and/or timeliness. For example, annual data published by the Australian Bureau of Statistics (ABS), such as the counts of Australian business entries and exits, provide annual point-in-time figures with breakdowns by range of employment size; however, these data cannot be used to establish marginal shifts in the composition of employment across categories of firms over time. Therefore, there is no way of knowing which firms grew from being small to medium-sized enterprises, medium-sized to large, and vice versa. Also, summary and descriptive statistics published by the ABS and other national agencies are useful for measuring the attainment of targets agreed to by the federal government and the States; however, they are inadequate for evaluating the specific programs designed to achieve the targets or assessing the policy interventions of the State based on its own priorities. To undertake this level of analysis, micro-data that covers both the firms and the employees are required.

Building a framework that can address a set of public research questions across a range of policy domains requires integration of the efforts of multiple State agencies. Given its central role in supporting cabinet committees, the DPC is in a unique position to convene and steer the efforts of these agencies.

The innovation was developed to deal with the issue of data not being available to conduct sub-national economic policy analysis. The problem was that the State government lacked the ability to target spending to maximise the outcomes of its economic development initiatives. The data that inform policy and program design are spread across multiple agencies and jurisdictions, which means policy design is fraught with uncertainty. The innovation addressed this problem by linking local State-held administrative data sets with national administrative data. This process led to the instantiation of a new data set (SA BLADE), the first of its kind in Australia where a State government (SAG) has linked business-related administrative data with Commonwealth data (Business Longitudinal Analytical Data Environment or BLADE) for the purpose of conducting evidence based economic policy for a sub-national unit. This in turn afforded the development of new methods for observing and analysing bespoke specialist industry sectors, innovation districts and State-based programs. The development of bespoke sectoral, district and program analytics has been made possible through new forms of inter-agency and institutional collaboration. Thus, the organisational boundaries and capacities of each party to work collaboratively have been tested. Also, the reduced level of specialist knowledge and technical expertise among policy units at the sub-national government level has made it necessary to reach out to academic partners to fill the gaps. Similarly, the need to demonstrate policy relevance and industry impact means that the academics need to reach out and work with policy units.

The creation of various forms of bespoke analytics was intended to fill a gap in the evidence base, as relevant data were needed to inform location-specific or place-based development strategies. If this evidence base and the methods for deriving them could be established, it would transform the way in which State government appraises past strategies and develops new initiatives in the future. The overall framework for expanding this evidence base, which would be generated by the innovation, would ultimately result in more prudent use of fiscal resources to support economic development.

Part 2: Literature review

The development and implementation of any IS is fundamentally a socio-technical endeavour. STS theory posits that the success of an innovation is not determined by technical excellence alone, but by the co-evolution and mutual adjustment of the technical components (hardware, software, data) and the social components (people, processes, culture, politics) (De Leoz & Petter, 2018).

The challenge of joining-up both social and technical systems is universal, but particularly prevalent in the public sector, where the use cases for adopting IS are based on their perceived public value (the result of maximising social benefits, while minimizing cost to the taxpayer), which is often more difficult to estimate, compared to the profit-motive in the private sector.

The literature on the use of big data, AI and machine learning (ML) in solving public issues has been growing (Athey, 2019; Mullainathan and Spiess, 2017; and Varian, 2014). ML has been used to improve tax administration (Aslett et al., 2024), provide predictive analytics in relation to social policy (Kleinberg et al., 2015), and improve equality and productivity using optimal tax policies through simulation (Zheng et al., 2020).

The literature on data linkages for economic research focuses on the various applications and usefulness of linked data in addressing issues related to poverty and income dependence (Jean et al., 2016; Meyer & Mittag, 2019; Sansone & Zhu, 2021), health (Young et al., 2018), housing (Messier, Elliott & Seitz, 2025) and tax policy questions (Gavin, 2021; Rivas & Crowley, 2018). It shows how combining different sets of data, including survey, administrative, satellite or geospatial, and macroeconomic statistics help to improve economic modelling and predictive power of existing data sets. This growth in the use of data linkage has come about because of the privacy-preserving record linkage techniques such as the five safes framework (Green & Ritchie, 2023).

The growing use of cross-sectoral data linkages has led to questions of how to govern them (Bartolomucci & Bartalucci, 2024). There is a growing body of work that focuses on the governance of what are termed "data collaboratives" (DCs) and the conditions necessary for DCs to persist beyond the pilot stage, which is quite rare (see for instance Susha, Janssen & Verhulst, 2017a and 2017b; Verhulst, 2023; Vernhulst, Young & Srinivasan, 2017; Kalkar & Alarcón, 2019). Much of the earlier work on the governance frameworks for DCs was done prior to the data revolution. This necessitates a review because, as Rujier (2021) states, DCs 'add socio-technical complexities', which 'may require new governance structures, processes and practices to ensure the proper working of collaboration.'

The growing complexity and lack of a universally acceptable governance model for handling cross-sectoral data linkages and DCs points to the possible use of an iterative model of co-evolution.

Action design research (ADR) is well-suited to navigating through the complexities of socio-technical issues as it directly ensures the co-evolution that is central to STS theory. The ADR literature suggests that getting the STS well-aligned is the best way to ensure innovation is adopted and diffused across its target audience. De Leoz and Petter (2018) state that

> the social impact of an (information technology or IT) artefact can be explained as the effect an artefact has on the

interpersonal relationships and interactions within an
environment at an individual, group, organisational, or
societal level of analysis (p. 156).

They also point out that social impacts are often unexpected
consequences that tend to be overlooked by IS research, as the
focus is often on the technical solutions. However, these
overlooked social consequences play a crucial role in improving
the adoption rate of artefacts generated.

Apart from looking at how the artefact serves its intended
purpose, they recommend that research should also examine the
artefact's fitness, and cite the work of Gill and Hevner (2013),
which describes fitness in terms of the artefact's 'ability to (1)
"survive at a high level of capacity over time" and (2) "to
reproduce – completely or in part – and evolve over successive
generations"' (Gill and Hevner, 2013, p. 3). The ability to achieve
the first is determined by the artefact's adaptability, and the
second is determined by the flexibility of the social structure. The
survival of an artefact requires that it have one of these two factors,
and when both are present, thriving occurs due to the ability of
both artefact and social structure to fit or adapt to each other. A
critical way to build this kind of adaptability and flexibility is
through "systematic reflection and competence building" as
Guertler, Kriz and Sick (2020, p. 387) emphasise by placing it at the
centre of their framework.

ADR focuses 'on the organisational context … and how this affects
the development and use of an artefact' (Peffers, Tuunanen &
Niehaves, 2018, p. 134). It is an integration of the action research
paradigm with design science research (DSR), in that it is 'a
research method for generating prescriptive design knowledge
through building and evaluating ensemble IT artefacts in an
organisational setting' (Sein et al., 2011, p. 39).

According to van Aken, Chandrasekaran and Halman (2016, p. 1),
DSR is 'a research strategy, aimed at knowledge that can be used
in an instrumental way to design and implement actions, processes
or systems to achieve desired outcomes in practice'. Gregor and
Hevner (2013) state

In IS, DSR involves the construction of a wide range of socio-technical artifacts such as decision support systems, modelling tools, governance strategies, methods for IS evaluation, and IS change interventions (p. 337).

By providing a detailed, reflective account of the three ADR cycles that shaped the SA BLADE project in this paper, we move beyond a simple description of the innovation to provide an analysis of the process of its success. This allows us to answer the research question posed in the introduction: How does a collaborative, multi-cycle ADR approach facilitate the co-evolution of STS to ensure the successful implementation and adoption of a sub-national linked data asset?

Part 3: Innovation, process & stakeholders

SA BLADE is the first type of data integration of its kind in Australia where data from a State government are linked to Commonwealth data for the purpose of developing evidence-based advice for a sub-national government.

The State component of the data set comes from the South Australian Business Research Environment (SABRE). SABRE is a data-sharing arrangement that has been forged by 13 agencies. The spine of this data set consists of the client database of Return to Work SA (RTWSA). RTWSA is a mandatory State-based worker's compensation insurance scheme that covers employers whose total remuneration costs exceed $13,423 in total over a year as of 2021-22 (Return To Work SA, 2021). Under the *Return to Work Act 2014 (SA)*, employers are required to insure their workers for lost compensation due to work-related injuries. The low compensation threshold for mandatory inclusion in the scheme means that almost all employing firms in the State are covered. Data from RTWSA is generally considered to be of better quality than similar data collected by its counterparts in other jurisdictions. For this reason, the ABS became interested in using it as part of the pilot to improve location data (Australian Bureau of Statistics, 2020).

Due to the incentives to accurately report the nature of work and number of workers at each business location, the RTWSA data became useful in disaggregating the business activities of firms with multiple State operations. While the federal tax office data report one address per corporate entity, the RTWSA data may contain multiple locations for an enterprise or enterprise group, with accompanying compensation levels associated with each location, making it an ideal data set for studying sub-national employment activity via the remuneration levels reported for insurance purposes.

BLADE combines the financial and employment information of firms collected by the Australian Taxation Office (ATO) with survey data on firm characteristics, finances and productivity collected by the ABS. BLADE allows for the analysis of businesses over time and the testing of hypotheses regarding conditions that help drive performance, innovation, job creation, competitiveness and productivity. It has been used by the Commonwealth to estimate the impact of its policies across Australia. The feasibility of linking the RTWSA and other SA-based data sets with BLADE was proven by the pilot project in 2018–2020. This pilot was undertaken by the SA Government and led by the DPC and ABS. The ABS had indicated strong support for this pilot project as a nation-leading example to demonstrate what may be possible from linking State data to the (already powerful) linked Commonwealth data in BLADE (Australian Bureau of Statistics, 2020).

Linking business data from the State and the Commonwealth creates a whole new data set. The strength of SA BLADE comes from being able to disaggregate the activities of national firms into their regional components.

In addition to these business-related micro-data, the ABS has also made individual-related data accessible through the Person-Level Integrated Data Asset (PLIDA). In its core product, PLIDA brings together personal income tax, social security benefit and Medicare-related data. These form the triple spine of PLIDA, and data that may be bolted on are the census, immigration, higher education, apprenticeship and Australian health survey data. The linking of employer and employee data in the SA BLADE innovation project expanded the possible uses of the data to include policy research covering demographic and social issues as they relate to intellectual, organisational and human capital formation within the State. Also, the firm and founder characteristics and behaviours of high-performing firms may be analysed, and the impact of policy drivers that are aimed at both the individual and firm levels can be tested and evaluated. In addition, the drivers of workforce and migration patterns within a State economy can be studied. The SA BLADE-PLIDA linkage can shed light on the occupational and demographic composition of the various sectors across precincts and regions, and it can provide information about the workforce needs of growth sectors, including those being targeted by the State government. The three main components of SA BLADE that were discussed above are presented in Figure 1 below.

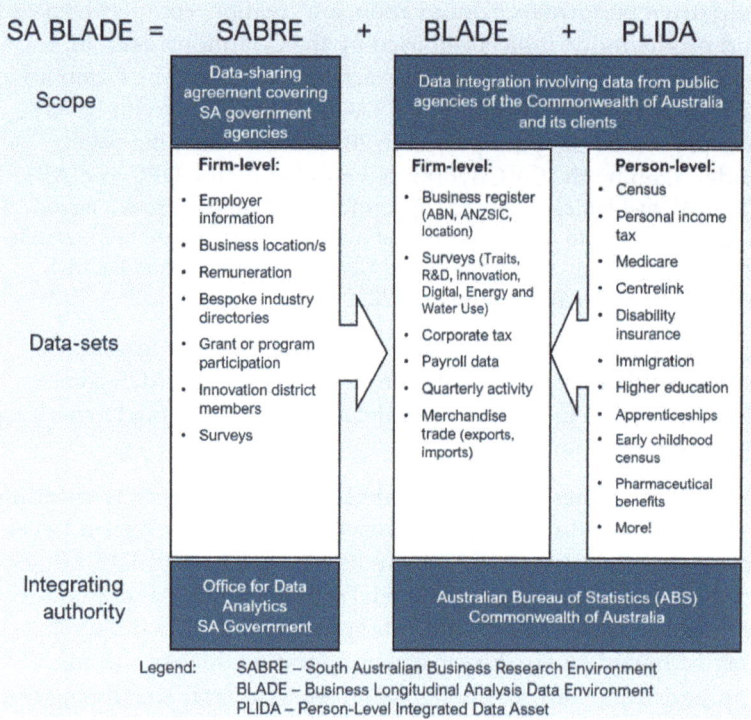

Figure 1: Components of SA BLADE
Source: Authors' own construction

The innovation was gradually developed over three successive ADR cycles. The first cycle consisted of the pilot phase and took place over three years from 2017/18- 2019/20. The aim of the first cycle was to test the feasibility and usefulness of linking State-based data with Commonwealth data. The State-held administrative data covered business location (from RTWSA) and compensation information (RTWSA and Revenue SA). It also covered program data from the State's Department of State Development (DSD).

The second cycle involved more data sets (10 compared to three in the first round) and a new process that involved a State-based linkage through the DPC's Office for Data Analytics (ODA) to create a new enduring IS artefact, the South Australian Business Research Environment (SABRE). SABRE was then integrated into BLADE to create SA BLADE 2.0. This additional step was introduced to manage the increasing complexity of handling more data sets. It removed the need for the ABS to deal multilaterally with various agencies as it was able to simply deal bilaterally with the DPC. The new arrangements were streamlined by making ODA the single point of contact. The second cycle added bespoke sectoral data and an experimental data set from one of the State's innovation districts. It also involved a separate investigation using PLIDA data to study interstate migration patterns.

The third cycle involved an update of the bespoke sectoral data and program data. It also included expanded coverage of innovation district data to two State-run precincts: Lot Fourteen in the Adelaide central business district, the site of the old Royal Adelaide Hospital, and Tonsley Innovation District, the site of the defunct Mitsubishi auto manufacturing plant. The third round also included linking SA BLADE with PLIDA.

The study commenced after the close of the first round and ended following the conclusion of the third. The focus group discussions were conducted as the second round was nearing completion and as the third round was underway. Qualitative client surveys were also conducted at the conclusion of the third round. The authors' reflections cover the length of the project, from conception all the way to the commencement of the third round.

The innovation in this study is a process and technology innovation that will ultimately lead to organisational and social innovation. The first part of the innovation project involved improving the existing administrative data frame by linking national and sub-national data sets. This step would culminate in the second part, which involved using this enduring data asset for collaborative policy research that could lead to the development or re-design of economic plans, programs and policies to benefit the local economy. The innovation can also be characterised as an incremental or sustaining innovation, in that it seeks to improve on the information contained in existing data assets, building on lessons learnt from a pilot phase that began in 2018. It can also be categorised as an open innovation, in that the State government sought to involve academics and research interns to co-design the innovation, participate in the process of transforming data into new knowledge for use by policy makers. This was done through several workshops facilitated during the pilot phase and a community of practice during subsequent phases.

The key influencers of the project fall into two categories. The first are the data users and adopters of the innovation. These are the senior State government ministers, executives, managers and officers responsible for the development of economic or industry policy. They are based in the central agencies, like the DPC and the Department of Treasury and Finance (DTF), as well as economic line agencies, such as DSD. Finally, the ABS also became a user of the data shared with them.

The second group consists of the data custodians and process owners. They include the ABS, which acts on behalf of the owners of the data found in BLADE from the Commonwealth side and is also the process owner of the data integration. Return to Work SA, Revenue SA and the other State agencies that contributed their data are the owners of the data and contributors from the State side.

There are five groups of enablers. The first group is the data integrators. They are the integrating authorities at the Commonwealth and State government levels that are empowered by their respective statutes. The ABS and ODA of SA are in this group. The second group of enablers is the research, business advisory and knowledge sector. In this category are private consulting firms, the universities, higher education and research units and parties. The third group consists of interstate and intergovernmental agencies, such as the Department of Industry, Science, and Resources (DISR), Commonwealth of Australia, and "Team Australia", which is a working group consisting of a loose collection of Commonwealth and State government representatives that meet regularly to discuss start-up, entrepreneurial and innovation policies and metrics. The fourth group consists of policy advocates and third-party organisations, including industry groups that cover the creative sector that participated in an industry mapping and measurement exercise. The final group consists of members of the precincts and the start-up ecosystem.

Part 4: Framing of success and results

SA BLADE is a contextually adaptive project that lends itself to an ADR mode of enquiry as discussed in the literature review in Part 2. The ADR framework involves repeated cycles of developing, evaluating and refining the IS artefacts by collecting lessons and insights from sequential iterations of the innovation project cycle. As Waterman et al. (2007) argue, action research is useful in investigations such as this that involve innovations for its 'propensity to innovate, diffuse innovations, and research innovation diffusion simultaneously' (p. 374). They say that unlike conventional research methods, such as the experimental or quasi-experimental methods where researchers 'seek to observe and report without being directly involved in the process of diffusing innovations', action research combines both innovation and research, because 'the approach recognizes that it is very difficult to separate innovation diffusion research from the innovation development and adoption process itself' (Waterman et al., 2007, p.

374). Another distinguishing feature of action research, according to them, is its facilitation of reflection and research, which is what distinguishes it from a standard change management process.

The research paradigm associated with action research is predominantly subjective and constructivist in nature, while also making use of a critical realist and interpretative research paradigm (Perry, 2013). The critical realism comes by seeking to explain emergent phenomena like policy frameworks and design artefacts observed through the innovation project. In SA BLADE, the IS artefacts would consist of policy relevant products, namely reports, analytical tools, data frames and systems, spinoff projects and insight briefs that rely on the innovation. The success of a design is based on subjective perceptions of their usefulness, quality and efficacy (Hevner et al., 2004). The artefacts generated are filtered through a subjective lens by their users to determine whether they add value to their individual and collective endeavours. The ADR methods allow participants to interpret what these artefacts mean to them, individually and as a group, after repeatedly interacting with them. Adoption and diffusion will depend on a net positive assessment by groups of users and stakeholders. ADR requires qualitative approaches that integrate reflective practice on the part of the researcher and participants. Reflection is what affords the construction of collective meaning, and ADR seeks to capture these subjective assessments through the harvesting of the key insights and learning gained by participants from their involvement in the innovation project.

One clear sign of adoption and diffusion is if the innovation changes the way individuals, groups and organisations interact within a given social context or environment (de Leoz & Petter, 2018). Evidence for this can be gleaned from statements found in public documents of the organisation sponsoring the project, such as websites, strategic plans and budget papers. They can also be determined through the personal observations of participants that have been shared within a community of practice and collectively interpreted through that process. ADR that incorporates a portfolio of evidence facilitates the compilation and interpretation of these different bits of evidence. An action research study calls for at least two successive rounds of development, implementation and reflection (Perry, 2013). Each iteration involves a cycle of action learning as defined by Revans (2011). The iterative process allows different approaches to the intervention to be tested, modified and tested again. In this study, collaborative planning sessions followed by implementation and reflective practice, informed by focus groups, were employed to draw lessons and further refine the collection and use of data (Santos, 2023). This was followed up with client surveys. The following applications of the linked data were found to be most useful in performing sub-national economic policy analysis by participants of the project.

Local impacts of adverse events

Estimating the financial impact of natural and man-made shocks on specific sub-national economies for better targeting of assistance was one of the benefits of SA BLADE. Following the SA bushfires of December 2019, State agencies, tasked with estimating the extent of the impact of this event on businesses in one affected region (Kangaroo Island), could provide an accurate picture of the financial assistance needed by business owners to rebuild the local economy. This was afforded by sharing of worker compensation insurance data that came about because of SA BLADE.

The mandatory nature of worker compensation insurance in the State meant that remuneration data for each location of operation had to be accurately reported by employers for proper coverage of workers. The sharing of data through the project informed decisions around the provision of assistance requested by businesses in the area. One focus group participant remarked that this was "the most useful application of this data." Another observed that this demonstrated administrative data held by the State had "a strategic use." The ABS subsequently used learnings from SA BLADE to rollout collection and integration of similar data from other jurisdictions to improve the business location information in BLADE.

Bespoke industry and ecosystem profiling

Closer proximity enables sub-national authorities to have better information regarding local industries to tailor public services. A challenge comes when monitoring growth of some specialist, cross-cutting sectors, which do not fall neatly under standard industry classification systems used to measure national accounts. Examples of this include defence and creative industries, which were identified and measured using SA BLADE by linking business directories supplied by industry associations, augmented with web-based desk research. The ABS used data shared through SA BLADE to help generate experimental estimates of the defence sector (Australian Bureau of Statistics, 2024), while the SA Government published a report on businesses in the creative industries (Government of South Australia, 2021) and knowledge and technology-intensive industries in South Australia with a special focus on its innovation districts that included start-up ecosystems (Santos et al., 2025). This allowed industry associations and State government to justify assistance provided to these sectors.

One participant said, SA BLADE "allowed us to respond to an industry recommendation and [develop a] creative industries strategy, which was what industry wanted to see". Firms could also be identified through web scraping to look for "technology hotspots" and identify innovative technology-based start-ups using data shared through the DC and from third party platforms (Han et al., 2023). The Australian Business Numbers (ABNs) of firms discovered from this process could then be searched and used to link to BLADE to estimate their aggregate sectoral financial performance. The combination of user-generated content in conjunction with administrative data allowed the State to have a better handle on bespoke sectors and ecosystems, including where activity is concentrated, with a view to shaping place-based interventions to support them.

Program evaluations

One of the biggest challenges facing sub-national political units when it comes to program monitoring and evaluation is the lack of data, especially where national governments administer the tax system, as in Australia. The absence of reliable financial data covering periods before and after interventions, make post hoc evaluations subjective, being heavily reliant on self-reporting by recipients. It also prevents construction of counter-factual evidence to show the marginal contribution of the program compared to an untreated control group.

SA BLADE created a shared program data frame that allowed the consistent collection of standard data items across various policy areas. These areas included innovation and commercialisation, trade promotion and investment attraction, energy and mining, primary industries and regional development, business investor migration, and green industry development. This allowed several policy initiatives to be rigorously evaluated once their program data was linked to BLADE (Rahman, Ambaw & Santos, 2023a, 2023b and 2023c).

One survey participant stated that, SA BLADE "added to our knowledge base of business outcomes to assist with briefing senior managers and the Minister ... [The result] will assist with designing future State business policies". Another remarked

> The evaluation was a very good piece of work, highlighting quantitatively [through difference in differences statistical approach] the contribution of the program to firm operational performance

and that the "ex-post evaluation will be a useful reference in any future related program design". A rigorous approach to program evaluation made possible by SA BLADE is yielding policy prescriptions based on what works, better criteria for screening participants, and improved focus for public investment.

Three levels of assessment

To evaluate the goodness of fit of the SA BLADE model within the context that it was used, evidence is collected at three levels. At each level, several artefacts are cited along with the accompanying reflections and realisations drawn by the study's participants. At the technical level, we look at three aspects: technical feasibility, data quality and privacy/security. Some of the insights are presented in Table 1 below. At the organizational level, we evaluated the innovation and its artefacts based on utility, relevance and adoption. These are shown in Table 2. At the social level, we uncover whether the new data innovation was meeting the needs of external stakeholders, whether it was changing attitudes and ways of interacting within and across State agencies and beyond. Table 3 contains what we found.

Criteria	Artefacts	Reflections/realisations
Feasibility	SA BLADE linkages (ver. 1 to 3)	High linkage rates of SA Government datasets to BLADE (each had a match rate above 90%) spoke to the high quality of ReturnToWorkSA, RevenueSA and business program data - Australian Bureau of Statistics, 2020
Data quality	ABS pilot evaluation study	The integrated dataset was a valuable information asset that expanded the evidence base available to inform SA Government's economic development policy…The benefits that these insights provide will continue to be realised beyond the pilot phase. – Australian Bureau of Statistics, 2020
Data privacy / security	ABS case study online	The pilot project underwent a rigorous assessment and approval process, managed by the ABS, and was overseen by a Steering Group of senior officers from the SA Government, the ABS and the Department of Industry, Science, Energy and Resources … Authorised researchers were granted access to de-identified microdata for policy analysis, research and statistical purposes. – Australian Bureau of Statistics, 2020

Table 1: Technical level of analysis

Criteria	Artefacts	Reflections/realisations
Usefulness/ insightfulness	New business location module in BLADE	The integrated dataset provided better identification of employing firms operating in SA and improved utility for regional analysis ... Relying on location information available in BLADE would have biased job counts toward States where large, complex businesses are registered for tax purposes. – Australian Bureau of Statistics, 2020 "It's basically enabling a better and more detailed understanding of things like opportunities, challenges, performance, contribution to the economy." – Focus group participant "SA BLADE was so groundbreaking" in that "it allowed greater accuracy than we'd ever had, and when you hear people ... in different parts of Australia talk about ... (how) we just relied on old census data before ... that's when you realise what a powerful tool it actually is" – Focus group participant
Usefulness/ insightfulness	Evaluation study	"The evaluation was a very good piece of work, highlighting quantitatively the contribution of (the program) to firm operational performance." – Survey participant "Highlighting contribution of audit component (of programs is) insightful and leads to consideration of options for future program design." – Survey participant
	Regional profiles	"(T)he profiling pieces have been unexpectedly more useful than perhaps we were first contemplating. The most useful application of this data is probably the low-level geography information that you can get." – Focus group participant

Criteria	Artefacts	Reflections/realisations
Relevance (policy / knowledge)	Evaluation study	"This was a valuable report … a welcome addition to our knowledge base." – Survey participant
	Policy review	"(T)he report will continue to be useful in scoping and assessing further initiatives for that portfolio and for the subsequent phases of the small business growth strategy" – Internal agency report
Adoption and diffusion (replicability, scalability)	Industry dashboard	"We will definitely be using this in an ongoing manner to inform many investment and policy decisions (for) broader industry development." – Focus group participant
	Pure research papers	"I can see enormous opportunities for doing some interesting academic stuff with the data" (and) "hopefully in the future to get other types of data that's confidential and that we can use it in an effective way" – Focus group participant
	Decision making tool	"You can think about using this to inform policy, design and development. You can think about it in terms of program evaluation, but also profil(ing), understanding, you know the nature of businesses in particular regions, and when something happens, if there needs to be a response, (and) you know what might the scale of that response needs to be." – Focus group participant

Table 2: Organizational level of analysis

Criteria	Artefacts	Reflections/realisations
Knowledge (frameworks, models)	Networks, social capital	SA BLADE was "a project that brings us together across government with a sort of shared purpose and a shared understanding". "It's bringing people to the table. It's building bridges … it's connecting with people in the research community." - Focus group participants
Learning (what works?)	Mindset	"There was a lot of nervousness around sharing data, and a lot of concern about how it (might) be used or misused. Having the hook of being able to link into the Commonwealth data provides some early answers that we just couldn't (get) with (our) own data sets has opened up the idea that some of our previously tightly held admin data does have a strategic use." - Focus group participant
Impact / Fitness (policies, outcomes, change)	Club goods	SA BLADE has "actually allowed us to respond to an industry recommendation and creative industries strategy, which was what industry wanted to see." – Survey participant

Table 3: Social level of analysis

Part 5: Discussion

The results section demonstrated that the success of SA BLADE was not a simple matter of technical execution; rather, it was the direct result of the ADR process.

This paper contributes to the literature in three ways. Firstly, on the governance of data collaboratives, the paper highlights the importance of co-evolution between social and technical systems through iterative problem solving which was a result of the ADR methodology. This fostered deep engagement through co-design and co-development by data end-users, IT developers, and data scientists, as an important factor of success.

As the literature indicates, very few data collaboratives persist beyond their pilot stages. SA BLADE is an instance where the data collaborative not only survived its pilot phase, but also gained increased usage and scale through repeated iterations, enduring the turnover of staff, senior management, and even changes of parties in power at both the State and Federal levels. It provides clear evidence supporting the thesis that deep participation and collaboration is an important ingredient to ensure the survivability and replicability of the data collaborative. The current literature only speaks of the need for stakeholder participation in governance, but SA BLADE included them in both governance and operational matters. In navigating the multi-agency data-sharing issues with ADR, SA BLADE demonstrated how initial nervousness was mitigated by involving data owners and policy end-users in the design and evaluation cycles. Secondly, the case as embodied in the paper extends the current use and application of data linkages to evaluating the impact of sectoral and place-based development programs aimed at firms and industries, and for start-up ecosystems and innovation districts. The current literature covers social, health and tax policy aimed at households and communities. Using linked administrative data and ML to predict where future new businesses and technologies might emerge and grow is yet to be fully explored. The findings hint at where and how to go about doing this.

Thirdly, the paper provides a theoretical contribution to ADR by demonstrating its efficacy as an overall framework for governing data collaboratives. This is very important in the age of digitalisation where many companies are striving to automate their decision making by adopting AI and ML in the management of their enterprise.

The paper strengthens the view that ADR can deepen participation and engagement in the design and implementation of IT and data-driven projects, an important condition for success. The adoption of AI and ML does not mean that human learning and participation are no longer needed. The case of SA BLADE as shown by the paper demonstrates that through action learning and action research social and technical systems can achieve a goodness of fit among users of data insight and the analysts that produce them, senior leaders and rank and file staffers in a mutually conforming and evolving loop of co-design and co-creation.

Wider implications of project & study

There are many ways to apply the know-how gained from SA BLADE as a model for using linked data for place-based economic development policy analysis.

As hinted above, succeeding rounds could include survey data collected from the State's innovation districts and start-up community (Alemayehu et al., 2023). Another plan is to construct a representative sample of the State's economy using SABRE. This sample will perform business surveys that gather information about their future intentions, managerial capacity, constraints to growth and perceived opportunities. The results could be linked to administrative data and used to test several propositions about the conditions that foster growth.

Data can be gathered from social media, geo-spatial and third-party platforms or the web in general to build on the research into technology hotspots and hubs of entrepreneurial activity already performed by the project. The businesses identified through sectoral mapping in the earlier rounds of the project could be used as the learning set for a ML algorithm to find similar businesses on the web. The ABNs of businesses discovered from this web-scraping exercise could then be linked to administrative data to estimate their financial and economic outputs.

Apart from economic grants provided by the Commonwealth and the State, councils also provide many incentives and collect a great deal of information. In addition, the utilities companies can provide rich data on water and energy usage rates. SA Water, for instance, has signed the SABRE agreement, making it possible for them to contribute data in the future. The SAG sold the office that handles land and property transactions with the condition that the data could be accessed by the State for policy research purposes. Data-sharing agreements could be forged around places of interest (e.g., regions of urban renewal and/or where large shipbuilding/defence contractors are based) to gain a more complete understanding of the local economic landscape and the spill over effects of place-based and district policies.

PLIDA provides another opportunity to link population, housing, education and health data between the Commonwealth and the States. A parallel PLIDA-based project called "Thriving SA" has been launched by the SA BLADE team to investigate the feasibility of doing this through the linkage keys supplied by the SA government data within the SA-NT Data Link (University of South Australia, 2022). It would be possible to tackle issues involving poverty, inequality, well-being and access to services through this effort. Some initial insight briefs and reports have already been generated through this (Yunren and Santos, 2023 and 2024)

As we saw in the case of the Robodebt scheme, using AI and ML for compliance purposes is fraught with risk. For purposes of ethics and risk management, the SA BLADE project scope veered away from managing service delivery in real-time and focused instead on policy design, appraisal and monitoring (although ML tools such as ordinary least squares, probit, and other econometric methods have long been used in econometric analysis of past policies).

It was our view that ethical and legal frameworks covering the use of AI for such purposes still have not caught up with the technology to deem them appropriate for handling compliance without human moderation or intervention. Instead, the project used Large Language Models such as Open AI to help analysts debug their codes and to fill-in their skills and knowledge gaps. For instance, they were used to augment desk research to cover blind spots, such as in identifying sectors that form part of supply chains of emerging green and technology industries that researchers were studying.

The use of AI and machine learning to help design industrial and innovation policy for instance is one area where the benefits could outweigh the potential risks associated with current practice. The perennial challenge faced by economic development planners is identifying the right places, firms, sectors or districts that need support, while avoiding the pitfalls of picking winners. The holy grail of economic policy is to identify with a large degree of confidence the recipients of assistance that would need and benefit from it. Criteria for directing such interventions are often set based on arbitrary lines of distinction without much evidence to support them. AI applied to linked administrative data could be used to find markers of future success in potential recipients that would not be found using standard econometric theory or models.

Conclusion

What does a successful data innovation project look like? In an age of data-enabled, AI-powered, automated algorithms, some would say, it is when machines are able to replace flawed humans in rational decision making. Many would answer this question by referring to the technical requirements. Is it technically feasible to perform? The innovation project covered in this study has shown that the technical feasibility is just the threshold issue. Of more lasting concern is the intelligibility, relevance and usefulness of the data that are linked, along with validation provided by humans in sense-testing the insights and decisions IS systems are recommending.

Given the primacy of user needs, it helps if users are involved in the design and implementation of the project. The success of SA BLADE can be attributed to its ability to address the intended needs of its users through co-design and co-development. Policy leaders who lived and breathed the programs were involved from start to finish, given agency and thus showed ownership.

As posited here, the decision-making and information systems co-evolved by mutually conforming to and mirroring each other. Data-driven policy does not mean that data projects should be scoped and designed mainly by data people. When policy teams with the questions and objectives relevant to the decisions that need to be made co-lead the data design process and analysis of the information, the result becomes more meaningful than if the data teams were left to their own devices.

A healthy interaction with thought leaders who push the boundaries of knowledge and are adept at pushing the boundaries of data to unlock patterns and uncover meaningful trends is also beneficial to policy makers. Often the world of academia is seen as an ivory tower that has no bearing on the real world, and practitioners often shy away from using their expertise knowing that their time horizons for delivery of output are often at odds. An important and significant lesson from the project in this study is that collaborative interaction allows policy and research to be mutually informed and directed. Thought leaders in academia have the models, theories and empirical methods to help frame and answer their theoretical questions, and practitioners have the relevant problems where these theories and approaches may be applied and tested.

The arbitrary chasm between knowledge-makers and user needs to be bridged to enhance the development of either side. Those willing and able to traverse the artificial barrier that separates them need to show an openness and flexibility of both discipline and practice to work with and learn from each other.

This study looked to trace and evaluate the development and adaptation of an innovative IS design artefact in shaping the policies and expanding the potential research agenda of the public agencies that adopted it. In doing so, it looked at the artefacts produced by the innovation itself. These artefacts reflect the external world that has been impacted and shaped by the project being studied. They provide evidence that the innovation is having some tangible, objective effect.

The reflections contained here are another important artefact of the study. They reflect the internal world of the researchers as shaped by observations and perceptions of the external world, interactions with other subjects of the study and personal thought processes as the innovation development and adaptation unfolded. The evidence for the footprint the innovation is leaving on these subjects is the subjective assessment of the meaning and significance that it is having on them.

The consistency of both outward and inward processes in this study is what supports its credibility and validity from an action design research perspective. When an innovation has both external and internal impacts that are valued and meaningful is how we know that it is likely to be adopted by an ever-growing number of users and supported by its beneficiaries.

We have found that any new context requires adaptation on the part of any system to achieve fitness for purpose. This gives rise to innovation as new elements responsive to local needs are added. Collaboration with partners and stakeholders, particularly the end-users, is key to fostering a socio-technical system of co-evolution that mutually informs the other process to avoid adverse unintended consequences. Action research and reflective practice help to foster learning and harness that learning for positive impact, continuous improvement and adaptation down the line.

Finally, from a theoretical perspective, this study contributes by extending the governance literature on data collaboratives to emphasise deep operational engagement, by broadening the application of linked administrative data to sectoral and place-based economic policies, and by strengthening action design research on socio-technical alignment. Together, these contributions highlight that meaningful data innovation requires not only technical feasibility but also co-evolution between human participation and technical systems.

Acknowledgements

The authors wish to thank the Department of the Premier and Cabinet (DPC), Government of South Australia, for allowing the study to be conducted by the lead researcher, and to the Australian Bureau of Statistics (ABS) for waiving the associated costs of data integration for the pilot featured here. Special thanks go to Dr Paul

Heithersay of the South Australian Government who chaired the initial pilot project for his leadership and sage advice, as well as to Alan Herning of the Economic Data Integration Team of the ABS for his dynamic and unwavering support.

References

Alemayehu, B., Steffens, P., Gordon, S. & Santos, E. C. S. (2023) *South Australian Innovation District (SAID) survey 2022: Preliminary findings*, The University of Adelaide in collaboration with the Department of the Premier and Cabinet, Government of South Australia.

Aslett, J., Hamilton, S., Gonzalez, I., Hadwick, D. & Hardy, M. A. (2024) Understanding artificial intelligence in tax and customs administration. *Technical Notes and Manuals* 2024 (6). International Monetary Fund. https://doi.org/10.5089/9798400290435.005.

Athey, S. (2019) The impact of machine learning on economics. In Agrawal, A., Gans J. & Goldfarb, A. (Eds) *The economics of artificial intelligence: An agenda*. University of Chicago Press, pp. 507–547.

Australian Bureau of Statistics (2020) *Use of integrated South Australian business data for economic analysis*, Available at https://www.abs.gov.au/about/data-services/data-integration/use-and-benefits/blade-case-studies#case-study-2-use-of-integrated-south-australian-business-data-for-economic-analyses.

Australian Bureau of Statistics (2024) *Australian Defence Industry Account, experimental estimates methodology*, Available at https://www.abs.gov.au/statistics/economy/national-accounts/australian-defence-industry-account-experimental-estimates/latest-release.

Bartolomucci, F. & Bartalucci, S. (2024) Designing data collaboratives' governance dimensions for long-term stability: an empirical analysis, *Data & Policy*, 6, p. e37. https://doi.org/10.1017/dap.2024.29.

Commonwealth of Australia (2023) *Royal Commission into the Robodebt Scheme* .Available at https://robodebt.royalcommission.gov.au/system/files/2023-09/rrc-accessible-full-report.PDF.

De Leoz, G. & Petter, S. (2018) Considering the social impacts of artefacts in information system design science research. *European Journal of Information Systems*, 27 (2), 154-170.

Gavin, E. (2021) How to collaborate effectively to improve data quality and use in revenue administration and official statistics. *IMF Working*

Paper No. 2021/05. Available at
https://doi.org/10.5089/9781513582863.061.

Gill, T. G. & Hevner, A. R. (2013) A fitness-utility model for design science research. *ACM Transactions on Management Information System*, 4 (2), 1-24. https://doi.org/10.1145/2499962.2499963.

Government of South Australia (2021) *Economic value of creative businesses in South Australia*, Available at https://www.dpc.sa.gov.au/responsibilities/economic-insight-and-evaluation/DIS-creative-industries-final-report.pdf.

Green, E. & Ritchie, F. (2023) The present and future of the Five Safes framework. *Journal of Privacy and Confidentiality*, 13 (2). https://doi.org/10.29012/jpc.831.

Gregor, S., & Hevner, A. R. (2013) Positioning and presenting design science research for maximum impact. *MIS Quarterly*, 37 (2), 337-355. https://doi-org.ezproxy.une.edu.au/10.1111/radm.12413.

Guertler, M., Kriz, A. & Sick, N. (2020) Encouraging and enabling action research in innovation management. *R & D Management*, 50 (3), pp. 380-395.

Han, H., Steffens, P., Mitchell, L. & Gordon, S. (2023) Adelaide's emerging technology heatmaps. *Economic Insight* 2023-01. The University of Adelaide in collaboration with the Department of the Premier and Cabinet, Government of South Australia.

Hevner, A., March, S., Park, J. & Ram, S. (2004) Design science in information systems research. *MIS Quarterly*, 28 (1), 75-105.

Jean, N., Burke, M., Xie, M., Davis, W. M., Lobell, D. B. & Ermon, S. (2016) Combining satellite imagery and machine learning to predict poverty. *Science*, 353 (6301), pp. 790–794. https://doi.org/10.1126/science.aaf7894.

Kalkar, U. & Alarcón, N. G. (2019) *Facilitating data flows through data collaboratives: A practical guide to designing valuable, accessible, and responsible data collaboratives*, Inter-American Development Bank. Available at: https://publications.iadb.org/publications/english/document/Facilitating-Data-Flows-through-Data-Collaboratives-A-Practical-Guide-to-Designing-Valuable-Accessible-and-Responsible-Data-Collaboratives.pdf.

Kleinberg, J., Ludwig, J., Mullainathan, S. & Obermeyer, Z. (2015) Prediction policy problems. *American Economic Review*, 105 (5), 491–495. https://doi.org/10.1257/aer.p20151023.

Messier, G., Elliott, S. & Seitz, D. (2025) Understanding housing and homelessness system access by linking administrative data. *arXiv preprint arXiv:2505.08743*. Available at https://doi.org/10.48550/arXiv.2505.08743.

Meyer, B. & Mittag, N. (2019) Using linked survey and administrative data to better measure income: Implications for poverty, program effectiveness, and holes in the safety net. *American Economic Journal: Applied Economics* 11 (2), 176- 204. https://doi.org/10.1257/app.20170478.

Mullainathan, S. & Spiess, J. (2017) Machine learning: An applied econometric approach. *Journal of Economic Perspectives*, 31 (2), 87–106. https://doi.org/10.1257/jep.31.2.87.

Peffers, K., Tuunanen, T. & Niehaves, B. (2018) Design science research genres: Introduction to the special issue on exemplars and criteria for applicable design science research. *European Journal of Information Systems*, 27 (2), 129-139.

Perry, C. (2013) *Efficient and effective research: A toolkit for research students and developing researchers*. AIB Publications.

Rahman, M., Ambaw, D. & Santos, E.C.S. (2023a) *Business investor migration visa evaluation report*. Government of South Australia.

Rahman, M., Ambaw, D. & Santos, E.C.S. (2023b) *Export acceleration grant evaluation report*. Government of South Australia.

Rahman, M., Ambaw, D. & Santos, E.C.S. (2023c) *SA energy efficiency program evaluation report*. Government of South Australia.

Return To Work SA. (2019, October) *Insurer statistics FY 2019*. Available at: https://www.rtwsa.com/about-us/return-to-work-scheme/scheme-statistics.

Return to Work SA. (2021) *Register for Cover*. Available at https://www.rtwsa.com/insurance/insurance-with-us/register.

Revans, R. W. (2011) *ABC of action learning*. Gower Publishing Limited.

Rivas, L. & Crowley, J. (2018) Using administrative data to enhance policymaking in developing countries: Tax data and the national accounts. *IMF Working Paper* No. 18/175. Available at: https://doi.org/10.5089/9781484371701.001.

Rujier, E. (2021) Designing and implementing data collaboratives: A governance perspective. *Government Information Quarterly* 38 (4). https://doi.org/10.1016/j.giq.2021.101612.

Sansone, D. & Zhu, A. (2021) Using machine learning to create an early warning system for welfare recipients. *IZA Discussion Paper* No. 14377, Available at SSRN https://ssrn.com/abstract=3851053.

Santos, E.C.S. (2023) *Creating a framework for conducting evidence-based policy for a sub-national economy using linked administrative data*. [Doctoral dissertation, University of New England]. Available from https://hdl.handle.net/1959.11/55586.

Santos, E.C.S., Yarashov, S., Huk, E. & Dang, T., (2025) *Knowledge and technology-intensive industries of South Australia: A study using administrative data*. Department of the Premier and Cabinet, Government of South Australia.

Sein, M. K., Henfridsson, O., Purao, S., Rossi, M. & Lindgren, R. (2011) Action design research, *MIS Quarterly*, 35 (1), 37-56.

Susha, I., Janssen, M. & Verhulst, S. (2017a) Data collaboratives as a new frontier of cross-sector partnerships in the age of open data: Taxonomy development. In Bui, T. X. & Sprague, R. (Eds.) *Proceedings of the 50th Annual Hawaii International Conference on System Sciences (HICSS 2017)*, University of Hawaii, vol. 2017-January, pp. 2691–2700. https://doi.org/10.24251/HICSS.2017.325.

Susha, I., Janssen, M. & Verhulst, S., (2017b) Data collaboratives as "bazaars"? *Transforming Government: People, Process and Policy*, 11 (1), 157–172. Available at https://www.sciencedirect.com/science/article/pii/S1750616617000142.

University of South Australia (2022) *SA-NT DataLink: Supporting health, social and economic research, education and policy in South Australia and Northern Territory*. Available at: https://www.santdatalink.org.au/.

Van Aken, J., Chandrasekaran, A. & Halman, J. (2016). Conducting and publishing design science research: Inaugural essay of the design science department of the Journal of Operations Management. *Journal of Operations Management*, 47-48 (1), 1-8.

Varian, H.R., 2014. Big data: New tricks for econometrics. *Journal of Economic Perspectives*, 28 (2), 3–28. https://doi.org/10.1257/jep.28.2.3.

Verhulst, S. (2023) *Data collaboratives: Enabling a healthy data economy through partnerships*, Center for the Governance of Change (CGC), IE University, Madrid.

Vernhulst, S., Young, A. & Srinivasan, P. (2017) Data collaboratives: Exchanging data to create public value across Latin America and the Caribbean. *Open Knowledge*, Inter-American Development Bank. Available at: https://blogs.iadb.org/conocimiento-abierto/en/data-collaboratives-exchanging-data-to-create-public-value-across-latin-america-and-the-caribbean/.

Waterman, H., Marshall, M., Noble, J., Davies, H., Walshe, K., Sheaff, R. & Elwyn, G. (2007) The role of action research in the investigation and diffusion of innovations in health care: The PRIDE project. *Qualitative Health Research*, 17 (3), 373-381.

Young, J., Borschmann, R., Camacho, X., Knight, J., Kouyoumdjian, F., Janjua, N., Atkinson, J. & Kinner, S. (2018) Linked data and inclusion health: Harmonised international data linkage to identify determinants of health inequalities. *International Journal of Population Data Science*, 3 (4). https://doi.org/10.23889/ijpds.v3i4.597.

Yunren, N. & Santos, E.C.S. (2023) Impact of household and neighbourhood income disparities on early childhood language and cognitive skills in SA. *Economic Insight 2023-02*. Department of the Premier and Cabinet, Government of South Australia.

Yunren, N. & Santos, E.C.S. (2024) *Do workers in South Australia's innovation districts earn more?* Department of the Premier and Cabinet, Government of South Australia.

Zheng, S., Trott, A., Srinivasa, S., Naik, N., Gruesbeck, M., Parkes, D. & Socher, R. 2020 The AI economist: Improving equality and productivity with AI-driven tax policies. *arXiv preprint arXiv:2004.13332*. Available at: https://arxiv.org/abs/2004.13332.

Biographies

Emmanuel Candido Soriente Santos, Ph.D.

Lead, Economic Insight and Evaluation, Department of the Premier and Cabinet, Government of South Australia (on leave)

emmanuelcssantos@gmail.com

Emmanuel is a public policy and international development adviser with over two decades of experience. He is the lead researcher and principal author of this study. He introduced action learning to the Philippine Civil Service Commission when he designed and developed a leadership program as part of an international development assistance project funded by the Australian Department of Foreign Affairs and Trade (DFAT). He led in facilitating the program with its maiden cohort in 2014/15. The Philippine government continues to benefit from its adoption to this day. Dr Santos' doctoral thesis in innovation incorporated action learning and action research.

ORCID: 0009-0000-9744-6321

Renato Andrin (Rene) Villano, Ph.D.

Professor and Chair, Research and Research Training Committee, UNE Business School - Faculty of Science, Agriculture, Business and Law; UNE Business School

rvillan2@une.edu.au

Rene Villano is a Professor of Economics at the UNE Business School, University of New England, Australia. A Distinguished Fellow of the Australian Agricultural and Resource Economics Society and a Senior Fellow of the Higher Education Academy, he has over 20 years of academic and research leadership in agricultural economics, applied econometrics, and development. He has participated in major multi-disciplinary research for development projects across Asia and Africa. Professor Villano has supervised over 80 graduate students and actively published in regional and international journals and contributes to several editorial and professional boards.

ORCID: 0000-0003-2581-6623

Jonathan Moss, Ph.D.

Senior Lecturer, Quantitative Economics, University of New England Business School

jonathan.moss@une.edu.au

Jonathan is an applied economist with a strong interest in integrating large spatio-temporal datasets with bioeconomic models to provide practical solutions to sustainability challenges in the agricultural and natural resource sectors. Recently, he has been working in cross-disciplinary and cross-institutional teams on the development of policies to increase farmer resilience under low carbon futures and the modelling and management of agricultural systems for sustainable production. He co-supervised the lead researcher in the study.

ORCID: 0000-0003-0462-8340

Benjamin Wilson, MEc (ANU)

Director, Economic and Environmental Policy, Department of the Premier and Cabinet, Government of South Australia

ben.wilson@sa.gov.au

Ben is a senior civil servant with more than two decades of experience working in the public sector both at the Federal and state governments. With a background in economics, Ben has provided advice in the areas of social, economic, infrastructure, regulatory, and fiscal policy. His contribution to the paper was to act as the direct supervisor to the lead researcher in the workplace, and as industry advisor in the design and implementation of the study.

# Action research and transformative learning in cross-sector healthcare collaboration

Heidi Lene Myglegård Andersen[1], John Andersen[2] & Ditte Høgsgaard[3]

Abstract

The growing number of elderly patients with multiple chronic conditions places increasing pressure on healthcare systems. Despite reforms, fragmented communication between hospitals, municipalities, and general practitioners still compromises continuity of care. This paper reports findings from the Danish Virtual 4-Meetings (V4M) project, an action research initiative aimed at improving transitional care for patients with multimorbidity through structured, digitally supported collaboration.

Grounded in participatory and critical action research, the study involved patients, relatives, hospital staff, municipal healthcare professionals, and general practitioners as co-researchers in iterative learning cycles. Data from workshops, virtual meetings, interviews, and field observations were thematically analyzed.

Findings show that the V4M model fostered shared learning, mutual understanding, and trust across sectors, transforming everyday tensions into opportunities for innovation. The study

---

[1] Faculty of Health Science, Copenhagen University College
[2] Department of People and Technology, Roskilde Universitet
[3] Department of Regional Health, University of Southern Denmark

demonstrates how digital facilitation, and participatory reflection can drive sustainable, cross-sectoral integration in healthcare.

Key words: Action research, transformative learning, cross-sector health care, virtual meetings

What is known about the topic?
Care for older adults with multimorbidity is often fragmented due to weak cross-sector communication. AR shows promise but is still underused in digitally supported, cross-sector collaboration.
What does this paper add?
It demonstrates how participatory AR and virtual meetings can strengthen cross-sector coordination. It also adds new insights into how collaborative learning unfolds across organizational boundaries.
Who will benefit from its content?
Professionals, managers, policy makers, and researchers working with coordination and transitional care.
What is the relevance to AL and AR scholars and practitioners?
• Shows how practice-based interventions generate multi-level learning • Demonstrates how daily routines can drive structural change • Offer insights into facilitation and insider roles

Received   July 2025   Reviewed   October 2025   Published   December 2025

Introduction

Across many healthcare systems, the growing number of elderly patients with multimorbidity presents both clinical and organizational challenges. These patients often navigate fragmented care pathways, where communication breakdowns, sectoral silos, and competing priorities undermine continuity and safety. Despite numerous reforms and pilot projects, the persistent structural divide between hospitals, municipalities, and general practice continues to limit integrated, patient-centred care. The result is often poor coordination and repeated hospital readmissions, leading to significant human and economic costs and considerable frustration among both patients and professionals. (Lemetti et al., 2021; Prior et al., 2024).

At the same time, top-down attempts to reform care delivery frequently fail to engage those who know the system best—the patients, relatives, and frontline practitioners. Without their involvement, system-level initiatives risk remaining abstract and disconnected from practice (Petiwala et al., 2021). This highlights the need for approaches that bring stakeholders together to co-create solutions grounded in everyday realities.

Action research (AR) offers such a framework. As Bradbury and Lifvergren (2016) argue, action research in healthcare can enable participants to act, learn, and change simultaneously, transforming both practice and understanding. By redistributing decision-making power and involving all actors as co-learners, AR fosters reflection, experimentation, and sustainable change. Yet, few studies have explored how participatory action research can be used to address sectoral transitions and the complexity of multimorbidity in a Danish context.

The Virtual 4-Meetings (V4M) project was designed to respond to this gap (Wentzer & Høgsgaard, 2022). Through structured, digitally supported meetings during hospitalization and discharge, the project brought together 1) patients and relatives, 2) hospital staff, 3) municipal providers, and 4) general practitioners to collaborate on care planning. This article examines how participatory and digitally mediated action research can strengthen cross-sector collaboration, facilitate collective learning, and support more coherent care for patients with complex conditions. The focus for this article is not to give a detailed description of project V4M but to describe the learning outcome of the action research process.

Specifically, we address the following research question:

> How can action research function as a catalyst for collaborative learning and system improvement across sectors in healthcare for patients with multimorbidity?

Theoretical framework

AR is widely recognized as a methodology that combines inquiry with action to achieve both practical improvements and theoretical insight. It is particularly suited to complex, cross-sectoral contexts where traditional research designs struggle to capture real-time learning and change (Coghlan & Brannick, 2014; Greenwood & Levin, 2007). In healthcare, AR has been shown to strengthen collaboration and quality improvement by engaging stakeholders as co-researchers in iterative cycles of planning, acting, observing, and reflecting (Bradbury & Lifvergren, 2016).

Within the broad AR field, participatory and critical traditions emphasize empowerment, co-creation, and democratization of knowledge (Brydon-Miller & Aragón, 2018). Rather than viewing patients and professionals as subjects of study, these approaches treat them as active partners in shaping the inquiry and the change process. Eikeland (2006) highlights *phronesis*, or practical wisdom emerging from shared experience, whereas Bacal Roij (2018) and Andersen (2015) emphasize that meaningful transformation arises through dialogical collaboration between researchers and practitioners.

Building on these perspectives, the study positions the researcher as an active participant and facilitator within the process, aligning with Coghlan and Brannick's (2014) notion of the "insider action researcher" rather than that of an external observer. The dual role of practitioner and inquirer enables continuous reflection on practice, supporting both personal and organizational learning.

John Dewey's (1938) pragmatist philosophy further enriches this perspective by framing learning as an active, social, and experimental process. Learning occurs when individuals engage with problematic situations, reflect on their experiences, and test new ways of acting. Building on this, the Danish action researcher Martin Frandsen (2018) conceptualizes participatory action research as a dynamic process of social learning, where tension points in everyday practice become catalysts for collective innovation.

These ideas are particularly relevant in healthcare, where structural divisions and power hierarchies often create friction. When such tensions are approached reflectively rather than defensively, they can stimulate new forms of collaboration and insight.

Wenger's (1998) theory of *communities of practice* provides an additional lens for understanding how knowledge is created and sustained across professional boundaries. Learning, in this view, is situated in social participation and shared meaning-making. In the context of cross-sector healthcare, virtual meeting spaces such as V4M can function as "micro-communities of practice," where professionals, patients, and relatives learn with and from each other through dialogue and experimentation.

Collectively, these theoretical perspectives form the conceptual foundation of the V4M study. They support an understanding of action research as both a methodological and relational framework, one that enables digital collaboration, shared reflection, and co-creation of knowledge across traditional sectoral boundaries.

Case description: The Virtual 4-Meetings (V4M) project

The V4M project was developed to improve coordination and care continuity for elderly patients with complex multimorbidity. It was carried out in Region Zealand, Denmark, through collaboration among hospitals, five municipalities, and general practitioners (GPs). The project's central idea was simple yet transformative: bringing together all key actors in the patient's care pathway — 1) patients and relatives, 2) hospital clinicians, 3) municipal providers, and 4) GPs — in a single, structured virtual meeting to plan admission and discharge together.

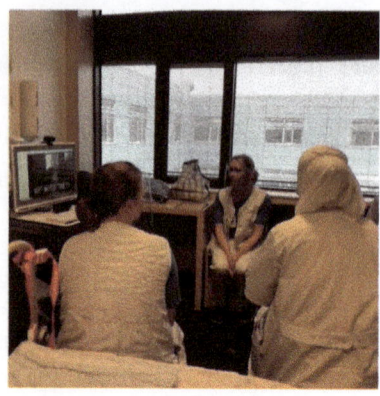

Figure 1. Picture of a video meeting:

The platform enabled real-time coordination, allowing professionals and patients to participate regardless of location. Hospital staff and patients were physically together at the bedside, while municipal staff and GP´s joined online. Meetings lasted a maximum of 30 minutes, and the meetings are patient-centred, guided by questions like "What is most important to you?" and "What expectations do you have for admission and discharge?" This collaborative approach ensured that concerns were identified, and a joint care plan was developed, tailored to the patient's needs. A patient expressed "*It gave me a sense of safety that they jointly agreed on treatment, aids and care after discharge*".

The aim was to move beyond one-way information exchange toward genuine dialogue, where all voices could be heard and collective decisions made. This dialogical structure addressed long-standing problems of fragmented communication and misaligned expectations across sectors.

Methodology

The V4M project was conducted as a participatory action research study combining insider inquiry, collaborative learning, and digital facilitation. The methodology followed the classical AR cycle of *planning, acting, observing,* and *reflecting* (Coghlan &

Brannick, 2014), enriched by participatory and critical perspectives (Brydon-Miller & Aragón, 2018; Eikeland, 2006).

Figure 2 below illustrates the cyclical nature of the action research process, based on the framework proposed by Coghlan and Brannick (2014). This framework served as the methodological foundation for the overall project and was applied consistently across all work packages.

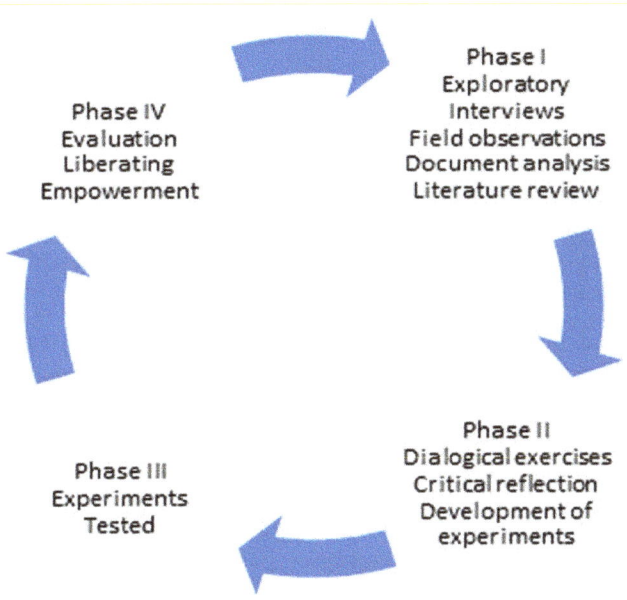

Figure 2. Action research process after Coghlan and Brannick (2014)

The overall project was divided into six work packages. Each work package represents a distinct component of the project, and all were developed through a participatory action research process:

- WP 1: Digital competence development for health professionals (www.virtuelle-konsultationer.dk; Mathiesen et.al., 2024).

- WP 2: Screening tools for patient inclusion (Christensen et al., 2024)
- WP 3: Health professionals' assessment and experience: An external evaluation (Wentzer, Risbjerg & Rayce, 2025)
- WP 4: Implementation and evaluation of V4M.
- WP 5: Evaluation of patient experiences
- WP 6: Addressing patient concerns about mental and physical movement post-hospitalization.

This article focuses on WP4 describing the projects overarching learning and action research process.

The project's organizational structure was the following:

- Action Research Group: Comprised senior action researchers, teaching researchers, research assistants (all nurses), and a PhD student specializing in multimorbidity[4]. Participated in the whole research project.
- Co-action researchers: Included patients, relatives, GPs, managers, doctors, nurses, therapists, and researchers from both hospital and municipal settings participating in the different work packages, video meetings, the organizational meetings and workshops.
- Steering Group: Included management from hospitals, five municipalities, the Association of General Practitioners in Region Zealand, and patient representatives. Participated in ongoing meetings and workshops.
- Advisory Board: Cross-disciplinary, with patient representatives, healthcare professionals, and senior

---

[4] Heidi Myglegård Andersen, Associate Professor, Ph.D., MPP, MA, RN: Ditte Høgsgaard Assistant professor, Ph.D., MPP, MScN, RN: Bettan Bagger Associate Professor, Ph.d. RN: Mai-Britt Hägi-Pedersen Assistant Professor PhD, RN: Marianne Kjestrup Research assistant, RN: Astrid Viôbjørg, Research Assistant, RN. Jon André Christensen Ph.d.fellow, cand.scient.

researchers, providing guidance on planning, implementation, and evaluation. Participated in ongoing meetings and workshops.

- Coordinator Group: Focused on identifying and refining organizational barriers, using the Future Workshop method. Participated in ongoing meetings and workshops.

Data generation.

Multiple qualitative data sources were combined to capture the complexity of the project.

Field observation, semi-structures interviews which were analyzed thematically following Braun and Clarke´s (2021) six-phase framework. Workshops were inspired by the Future Workshop (Jungk & Müllert, 1987).

Building on this foundation, the role of the action researchers was critical in facilitating the participatory process, bridging theory and practice, and fostering reflective learning among participants. The following section explores how the action researchers navigated different research traditions to support collaborative knowledge production, systemic change, and anchoring.

The role of the action researcher

Action research involves multiple roles and activities, emphasizing participation, context-dependent knowledge, and change (Brydon-Miller & Aragón, 2018). Coghlan and Brannick (2014) describe *insider action research*, where the researcher is embedded in the organization and acts simultaneously as practitioner and inquirer through iterative cycles of planning, acting, observing, and reflecting.

Critical and participatory approaches, as outlined by Brydon-Miller & Coghlan (2014), stress the collaborative creation of knowledge with stakeholders to promote empowerment and systemic change. Research thus becomes both a means of generating knowledge and a political act, linking reflection and action with structural critique. Bacal Roij and Eikeland particularly underline the role of phronesis, understood as practical wisdom that grows out of collective learning. They also emphasize the importance of dialogical and balanced relationships between researchers and practitioners. (Eikeland, 2006; Andersen, 2015; Bacal Roij, 2018).

The V4M project integrated these perspectives. The action researcher worked across hospital and municipal settings to connect theory and practice through co-creation workshops, strategic dialogues, and participation in cross-sectoral forums. These engagements helped translate project learning into concrete organizational changes, align stakeholders, and support the integration of new routines. The researcher acted not as an external evaluator but as a collaborative partner facilitating reflection and driving change from within.

Building on Coghlan and Brannick's (2014) insider engagement and Eikeland's (2006) collaborative knowledge creation, the project balanced top-down structures with bottom-up innovation. Through workshops and close collaboration with leaders and frontline staff, the action researcher cultivated shared visions, identified practical steps for change, and fostered mutual understanding across sectors.

The project action research group carried out the study as insider researchers, all being both experienced nurses and experienced researchers. This dual role required ongoing reflexivity to balance facilitation and analysis. Reflective logs, collegial supervision, and regular debriefing meetings were used to examine our influence on interactions and interpretations. As facilitators, we sought to create open and non-hierarchical spaces, encouraging all voices to be heard. As researchers, we documented how these spaces contributed to shared understanding and change.

Ethical considerations in the role of the action researcher

Ethical approval was obtained from the participating institutions. Informed consent was collected from all participants, and confidentiality was maintained throughout. Particular attention was paid to power dynamics. Early observations revealed that physicians often dominated conversations, while patients and municipal staff hesitated to speak. To counter this, facilitators used structured turn-taking and explicit prompts to ensure balanced participation. This approach not only protected participant well-being but also deepened learning about inter-professional communication.

Given the participatory nature of the research and the involvement of vulnerable populations, such as elderly patients, the V4M project placed a strong emphasis on ethical considerations and safeguards throughout the study. As an example, the project began with action researchers conducting in-depth conversations with patients to explore their experiences and perspectives. Insights from these discussions were then brought into a subsequent workshop, where patients, relatives, and healthcare professionals came together. The primary aim was to stimulate reflection and learning among health professionals by ensuring that the patient voice was actively represented in the dialogue. This process not only fostered a culture of openness and mutual respect but also helped to balance power dynamics, enabling all participants to contribute meaningfully to the learning process. Moreover, the action researchers played a pivotal role in continuously monitoring, addressing and mediating power imbalances in workshops and throughout the project.

In addition to patients often having a limited voice in their care process, there is also an existing power imbalance between doctors and other healthcare professionals. To address this challenge and promote more balanced cross-sector communication, a dialogical communication model to be used in the virtual meetings was developed – the Circle Care model (Høgsgaard et al., 2025).

The Circle Care model was developed through an action research process. The Circle-Care model is based on five core principles: 1) C – Collaboration: Sharing knowledge, 2) I – Involvement: Addressing the care and treatment needs of patients and their relatives, 3) R – Relationships: Jointly managing the patient's cross-sectoral care pathway 4) CL – Clear Communication: Engaging in dialogue about concerns and expectations 5) E – Embrace: Focusing on shared goals and actions (Høgsgaard et al., 2025).

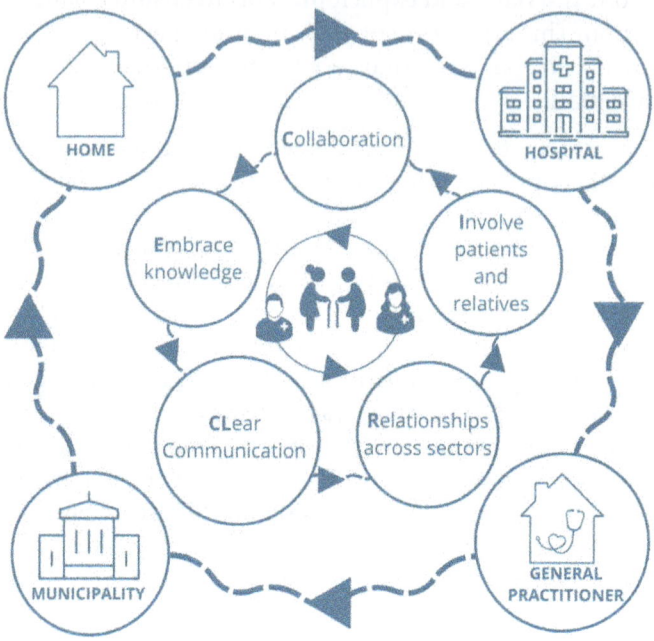

Figure 3. The Circle-Care model

The model responds directly to local needs in integrated care and emphasizes power balance by creating structured communication and inclusive spaces where patients, relatives, and frontline professionals can co-create solutions. This approach challenges traditional hierarchies and ensures that diverse voices, especially those often marginalized, are heard and valued in shaping coordinated care. (Høgsgaard et.al., 2025).

The integration of a clear structure with skilled facilitation enabled the dialogical communication model to support meetings where patients, relatives, hospital staff, municipal providers and general practitioners could participate equally and have their perspectives acknowledged.

By encouraging open dialogue and mutual knowledge exchange, the model fostered a collaborative environment for collective learning. The structured dialog enabled participants to learn from one another, dismantling traditional hierarchies and co-creating solutions tailored to local needs. For instance, a municipal nurse explained how a patient with severe lung disease could receive a nebulizer at home which was information previously unknown to hospital staff. The real-time dialogue facilitated timely medication adjustments and offered reassurance to the patient while enhancing mutual understanding among the professionals and their working contexts. Unlike one-way written communication, this dynamic interaction proved essential for promoting truly integrated care. By critically and actively addressing power dynamics, encouraging open dialogue, and involving all stakeholders, the Circle Care model served as a tool not only to generate safer care pathways but also to protect participant well-being and not the least to foster the emergence of new insights and collaborative knowledge.

One of the main barriers encountered in the project was the lack of consistent managerial support. While the V4M initiative proved meaningful for both patients and healthcare professionals, coordinating the meetings across sectors was time-consuming. As a research-based initiative, V4M was perceived as a *"could-do"* rather than a *"must-do"* task. Despite goodwill and positive intentions from leadership, there was limited structural commitment and prioritization.

To address this challenge, the action research team established collaboration with a managerial body known as the Regional Health Cluster, which was mandated to develop cross-sectoral solutions to organizational challenges in the care of older adults with multimorbidity. The researchers participated in planning meetings and presented both practical experiences and theoretical insights from the V4M project. This strategic engagement helped bridge the gap between research and organizational practice. As a result, V4M has now transitioned from a research initiative to an ongoing implementation process within the regional health system, supported by formal management structures.

Building on these ethical and participatory foundations, the following section examines how the V4M project generated learning through the action research process itself. Based on Martin Frandsen (2018) and John Dewey's (1938) pragmatist philosophy, learning is viewed as an experimental and social process that unfolds through action, reflection, and interaction among participants.

In this sense, the V4M project functioned as a living laboratory where patients, relatives, and professionals jointly explored new ways of collaborating. Analysis of data from workshops, interviews, and observations showed that learning unfolded across three interconnected levels: individual, social, and organizational. Across these levels, moments of tension, uncertainty, or disagreement served as productive catalysts for reflection and innovation, aligning with Frandsen's and Dewey's understanding of learning as rooted in lived experience and experimentation.

Learning through action research

John Dewey's pragmatist philosophy offers a foundation for understanding learning as both experiential and social. For Dewey (1938), learning arises when individuals engage with real-life problems, reflect on their experiences, and test new ways of acting. Knowledge is created through participation and experimentation, where reflection and action together transform experience into insight.

Building on this, the Danish action researcher Martin Frandsen (2018) extends Dewey's ideas into participatory contexts. He understands learning as a form of social experimentation, a collective process of exploring and interpreting everyday challenges. In action research, this happens when participants inquire into their own practices, co-create knowledge, and develop new understandings together.

From this perspective, learning is understood as both the method and the outcome of action research. When participants work together to address tension points in practice, where uncertainty or disagreement arises, they develop new insights and shared approaches. In doing so, individual experiences are transformed into collective and organizational learning.

According to Frandsen (2018), the prototypical model for participatory action research begins with everyday troubles and social injustices that create a need for collective action. The process moves from dialogue and partnership building to participatory inquiry, trust formation, and collective experimentation. Even when change efforts fail, learning still occurs through reflection on structural obstacles (Greenwood & Levin 2007).

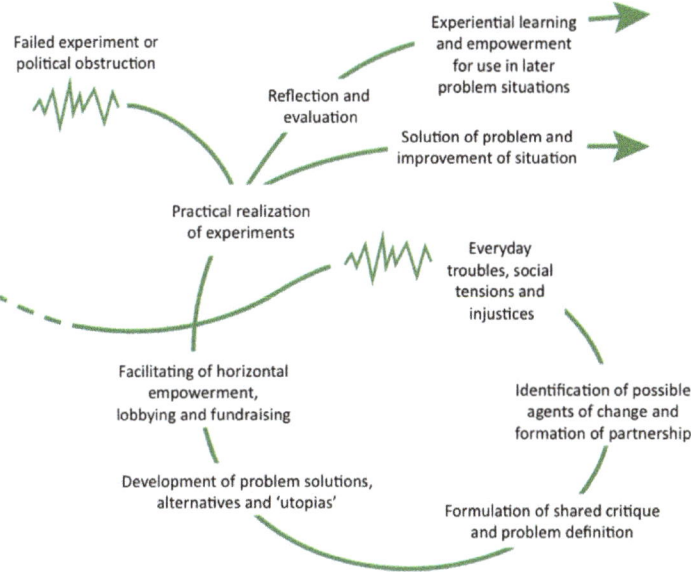

Figure 4. Frandsen´s (2018) prototypical model for participatory action research

The model begins with everyday troubles and social injustices that call for collective action. First, dialogue among affected actors identifies potential partners and a shared understanding of the problem. Next, participatory inquiry deepens knowledge, builds trust, and fosters mutual commitment. From this, collective action ideas and solutions are developed and tested experimentally. If solutions require institutional or political change, they can be promoted publicly and politically. Successful outcomes can be scaled up to other communities or levels, while failed experiments still provide valuable insights into structural barriers to change. The model also shows that tension points in collaboration can foster innovation and growth when handled constructively but can lead to stagnation if they are ignored (Frandsen, 2018). Frandsen emphasizes that tension points in collaborative projects act as catalysts for change. When they are handled constructively, they spark innovation; when they remain unresolved, they lead to stagnation.

In V4M, learning emerged from these tension points, including unclear roles, fragmented communication and differing assumptions about responsibility.

One hospital nurse explained: *"We often must discharge patients very quickly... I'm not always sure if the information I write in the electronic care plan reaches the right person."* A municipal nurse expressed: *"We don't always get the information we need... it's frustrating that we don't have the same access to patient information."* Facilitated exchanges in V4M enabled participants to "see each other's realities" and identify practical solutions.

Drawing on Frandsen (2018) and Dewey (1938), who view learning as both experimental and social, the V4M project showed that learning unfolded across three interconnected levels: the individual, the social and the organizational. Individual learning arose when professionals reflected on their own practices and assumptions. A hospital nurse noted: *"After hearing their side, I understood their challenges much better. We need to address that."* Patients and relatives also gained voice and agency. A patient said *"I felt like I was getting involved, and that the conversation was based on my wishes for the future. It was respected that I did not want dialysis treatment"*. And a relative said

We were very unsafe if dad could take care of himself when he got home. There was a good joint talk about what to do in the worst case. The nurses from the municipality highlighted that they are being able to come at short notice. We were still unsafe when he was discharged but it went well.

These reflections illustrate how individual learning emerged through dialogue and experience. For professionals, listening to others' perspectives prompted self-reflection and a re-evaluation of habitual practices. For patients and relatives, participation created new awareness of their own roles and agency in decision-making. Through these encounters, learning occurred as participants developed greater confidence, understanding, and capacity to act within their own spheres of responsibility.

Social learning developed through joint experimentation in V4M workshops and meetings.

Dialogical interaction fostered empathy, trust, and shared understanding, while simultaneously generating knowledge grounded in the specific context.

Reflection before, during, and after action, as emphasized by Dewey (1938) and Frandsen (2018), supported mutual adaptation and innovation. Through this process, communication patterns evolved, trust developed across sectors, and new shared practices emerged. Learning took place not in isolation but through interaction and the co-construction of meaning. Organisational learning occurred when insights became part of institutional routines. Hiring two research nurses to facilitate meetings increased participation and made collaboration meaningful. As one administrative manager observed: *"At first, I didn't see how these meetings would fit into our routines… The video meetings make a real difference. I realized we needed to support this more actively."*

Through iterative, reflective cycles, researchers and practitioners worked as equal partners in addressing everyday challenges. This process created "micro-practice communities" that fostered cross-sector trust and mutual understanding.

At the organizational level, V4M also demonstrated that digital facilitation and participatory reflection could generate systemic improvements. Initially, participation was voluntary and sporadic, but as evidence of patient benefit accumulated, leadership began to integrate V4M into standard care routines. The use of shared platforms and regular reflection workshops enabled innovations to spread across departments and municipalities, marking an important step toward what Wenger (1998) characterizes as sustained communities of practice.

Summary of Key Outcomes

Level of Learning	Example of Change	Resulting Impact
Individual	Professionals recognized the patient's lived experience and adjusted communication practices.	Improved patient inclusion and empathy
Social	Cross-sector participants developed shared understanding and trust.	Reduced fragmentation, stronger relationships
Organisational	V4M and Circle-Care model embedded in local routines.	Sustainable improvement in transitional care

Discussion

The findings from the V4M project demonstrate how action research can function as a catalyst for both learning and system transformation into complex healthcare settings. By engaging patients, relatives, and professionals as co-researchers in a structured, digitally supported process, the project generated new insights into how fragmented care can be bridged through dialogue and reflection.

A central finding concerns the productive role of tension points. What initially seemed to be barriers, such as miscommunication, competing priorities, or hierarchical norms, turned into catalysts for learning once they were made visible and discussed. This dynamic reflects Frandsen's (2018) and Dewey's (1938) argument that learning arises through engagement with problematic situations. When participants confronted misunderstandings about discharge procedures or patient needs, they collectively experimented with new solutions. Over time, this iterative process transformed relationships across sectors, building trust and shared ownership of patient outcomes. This shift from defensiveness to dialogue illustrates the transition from "doing coordination" to "learning coordination" which is key hallmark of action research in practice.

Action research provided the methodological and relational scaffolding for this transformation. Following Coghlan and Brannick's (2014) model, the insider researchers continuously cycled between inquiry and action, integrating learning from daily practice into broader organizational development. Unlike traditional top-down quality initiatives, V4M embedded improvement processes directly into the work routines of those most affected.

This approach aligns with Eikeland (2006), who emphasizes practical wisdom (phronesis) developed through collaborative reflection, and with Bradbury and Lifvergren (2016), who describe healthcare action research as "a way of learning our way forward". The Circle-Care model became a tangible expression of these principles, translating abstract values like participation and trust into structured communication practices that could be applied in daily care coordination.

Leadership support proved essential for moving from experimentation to sustainable integration. In line with international findings (Cornish et al., 2023), the project showed that cross-sector innovations cannot be maintained without strategic alignment and managerial endorsement. Initially, V4M depended on motivated individuals; later, it became part of organizational routines once ward and municipal leaders recognized its value in preventing adverse events and improving discharge quality.

The insider researchers served as a bridge, translating between the reflective domain of action research and the practical realities of healthcare management. By articulating the project's impact in terms of patient safety and efficiency, they secured long-term commitment from decision-makers. This underscores the dual responsibility of action researchers to facilitate learning *and* ensure institutional anchoring.

Digital facilitation emerged as both an enabler and a challenge. The online meetings made participation possible across geographic and institutional boundaries, aligning with recent studies on tele-collaboration in healthcare (Mathiesen et al., 2024). However, variations in digital literacy and infrastructure initially limited participation. The digital competence training developed in WP1 helped address this gap, but technology alone was not sufficient; relational trust and communication skills remained the true drivers of success.

Digital tools should therefore be regarded not as substitutes for human interaction but as extensions of dialogical practice, reflecting Dewey's view of tools as mediators of experience.

When used within an action research framework, technology can enhance reflexivity, inclusion, and continuity of care.

Ethical and Power Considerations

Ethical reflection was not a separate step in the project but an integral, ongoing process. Balancing power relations across professions and between professionals and patients requires constant attention. Structured dialogue formats, co-facilitation, and "voice rounds" helped ensure equity in participation. These practices reflect Brydon-Miller & Coghlan (2014), who argue that action research is both an epistemological and an ethical endeavour aimed at democratizing knowledge creation and questioning hierarchical norms.

In the V4M project, empowerment was experienced not as a slogan but as a lived process: patients' narratives influenced care planning, nurses negotiated professional boundaries, and physicians adjusted communication styles. This relational ethics of participation represents one of the most enduring impacts of the project.

Limitations and Reflections

Like most action research, this study was context-dependent and cannot claim universal generalizability. It was conducted within a specific Danish region, shaped by local structures and leadership cultures. The insider role of the researchers, while valuable for contextual insight, also posed risks of bias. Reflexive practices such as peer debriefing, journaling, and validation workshops were essential in addressing this challenge.

Nevertheless, the study provides transferable insights for other settings where fragmented care and digital transformation intersect. Its strength lies in illustrating *how* learning and change can be cultivated through participatory processes rather than imposed from above.

Conclusion

The V4M project contributes to the international discourse on action research in healthcare by showing how practice-based, participatory learning can strengthen cross-sector collaboration and create more integrated care pathways for older adults with multimorbidity. In line with global principles of community-based participatory research (Bradbury & Lifvergren, 2016), the project demonstrates that sustainable change emerges when patients, relatives, and professionals act as equal partners in developing solutions grounded in everyday practice.

V4M illustrates how iterative learning processes, facilitated by insider action researchers, can bridge theory and practice, enabling collective reflection and problem-solving across organizational boundaries. This supports international evidence that action research effectively drives improvement by embedding learning and change in participants' daily work (Coghlan & Brannick, 2014; Eikeland, 2006).

The project further shows how digital collaboration tools and structured meeting formats can translate action research principles into practice, creating "micro-practice communities" where trust, shared understanding, and local innovation thrive.

Achieving lasting change, however, depends on leadership support, digital competence, and continuous evaluation, challenges that are also reflected in international studies. The strength of action research lies in its capacity to create settings for shared learning and collaborative development, where everyone involved, from patients to leaders, can contribute to sustainable and integrated healthcare innovation.

Funding

The project has been funded by NOVO Nordic Foundation.

References

Andersen, H.L. (2015) Community health – *Sundhedsfremmestrategier og planlægning i lokalsamfund*. (*Community health – health strategies and planning in local communities*) PhD-dissertation. Roskilde; Roskilde University. Available from: http://rudar.ruc.dk/handle/1800/27071.

Bacal Roij, A. (2018) The pedagogical legacy of Dorothy Lee and Paulo Freire. In Misseyanni, A., Lytras, M.D., Papadopoulou, P. & Marouli, C. (Eds.) *Active learning strategies in higher education*. Bingley; Emerald Publishing Limited, pp 339-359 https://doi.org/10.1108/978-1-78714-487-320181015.

Bradbury, H. & Lifvergren, S. (2016) Action research healthcare: Focus on patients, improve quality, drive down costs. *Healthcare Management Forum*, 29 (6), 269–274. https://doi.org/10.1177/0840470416658905.

Braun, V. & Clarke, V. (2021) *Thematic analysis: A practical guide*. London; Sage.

Brydon-Miller, M. & Aragón, A.O. (2018) The 500 hats of the action researcher. In Bilfeld, A., Jørgensen, M.S., Andersen, J & Perry, K.A. (Eds.) *Den ufærdige fremtid – Aktionsforskningens potentialer og* udfordringer. (*The incomplete future: Potentials and challenges of action research*). Aalborg; Aalborg University Press. pp. 19-47.

Brydon-Miller, M. & Coghlan, D. (2014) The big picture: Implications and imperatives for the action research community from the SAGE encyclopedia of action research, *Action Research*, 12 (2), 224-233.

Christensen, J.A., Marcussen, M., Zabell, V., Bagger, B., Høgsgaard, D., Lundstrøm, S., Barrett, B.A., Mortensen, A., Frølich, A., Skou, S.T. & Berring, L.L. (2024) Stratification tools for elderly people with multimorbidity in need of integrated care: A scoping review. *Journal of Multimorbidity and Comorbidity*. https://journals.sagepub.com/doi/10.1177/26335565251357781.

Coghlan, D. & Brannick, T. (2014) *Doing action research in your own organization*. 4th edn. London; Sage.

Cornish, F., Nyutsem Breton, N., Moreno-Tabarez, U., Rua, M. & Hodgetts, D. (2023) Participatory action research: The missing bottom-up perspective. *Nature Reviews Methods Primers*, 3 (1). https://doi.org/10.1038/s43586-023-00214-1.

Dewey, J. (1938) *Experience and education*. New York; Macmillan.

Eikeland, O. (2006): Phrónêsis, Aristotle and action research. *International Journal of Action Research*, 2 (1), 5-53.

Frandsen, M. S. (2018): Sociale læreprocesser: John Deweys pragmatisme som udgangspunkt for aktionsforskning (Social learning processes: John Dewey's pragmatism as a basis for action research). In Bilfeldt, A., Jørgensen, M.S., Andersen, J. & Perry, K.A. (Eds.) *Den ufærdige fremtid – Aktionsforskningens potentialer og udfordringer* (*The incomplete future: Potentials and challenges of action research*). Aalborg; Aalborg University Press. pp. 69-99.

Greenwood, D.J. & Levin, M. (2007) *Introduction to action research: Social research for social change*. 2nd edn. Thousand Oaks, CA; Sage Publications.

Høgsgaard, D., Jensen, J.F., Andersen, H.L.M., Tang, L.H., Skou, S.T. & Simonÿ, C. (2025) A Circle-Care model in integrated care for patients with multimorbidity: An action research study. *Journal of Integrated Care*, 33 (2), pp. 182-196. https://doi.org/10.1108/JICA-09-2024-0053.

Jungk, R. & Müllert, N. (1987) *Future workshop – how to create desirable futures*. London; Institute for Social Invention.

Lemetti, T., Puukka, P., Stolt, M. & Suhonen, R. (2021) Nurse-to-nurse collaboration between nurses caring for older people in hospital and primary health care: A cross-sectional study. *Journal of Clinical Nursing* 30 (7-8), 1154-1167.

Mathiesen, L. M. W., Bagger, B., Høgsgaard, D., Nielsen, M. V., Gjedsig, S. S., Hägi-Pedersen, M-B. (2024) Education and training programs for health professionals' competence in virtual consultations: a scoping review protocol. *JBI Evidence Synthesis* 22 (12), 2618-2624. https://doi.org/10.11124/jbies-23-00285.

Petiwala, A., Lanford, D., Landers, G. & Minyard, K. (2021) Community voice in cross-sector alignment: Concepts and strategies from a scoping review of the health collaboration literature. *BMC Public Health* 21, 712.

Prior, A., Baymler, A.S., Lundberg, C.B., Vestergaard, U. & Utoft, N.B. (2024) *Research in the treatment of multiple chronic conditions in general practice*. Aarhus: Research Unit for General Practice, University of Aarhus.

Virtuelle-konsultationer.dk, n.d. *Virtuelle konsultationer (Virtual consultations)*. Available from Tilgængelig fra: https://www.virtuelle-konsultationer.dk [Accessed 10 July 2025].

Wenger, E. (1998) *Communities of practice: learning, meaning, and identity*. Cambridge; Cambridge University Press.

Wentzer, H. S. & Høgsgaard, D. (2022). *Tværsektorielle videomøder om den ustabile patient (Cross-sectoral video meetings about the unstable patient) A blueprint for conducting virtual four-party meetings on extended coordination, V4M, between hospitalized older patients, relatives, the hospital, the municipality and the patient´s doctor*. Copenhagen. VIVE. https://www.vive. – The Danish Center for Social Science Research. Available from Tværsektorielle videomøder om den ustabile patient - vive.dk.

Wentzer, H. S., Risbjerg, A. W. & Rayce, S. B. (2025). *Brugerperspektiver på 4-parts videomøder (V4M) med den multisyge patient (User perspectives on 4-party video meetings (V4M) involving the multi-morbid patient). Report on V4M 2.0: Evaluation of user perspective on virtual four-party meetings (V4M) with multimorbid patients*. Copenhagen. VIVE. The Danish Center for Social Science Research. Available from: https://www.vive.dk/da/udgivelser/brugerperspektiver-paa-4-parts-videomoeder-v4m-med-den-multisyge-patient-yxdr4jlz/

Biography

Heidi Lene Myglegård Andersen, Ph.D., MPP, MA, RN. Senior Reseacher. Copenhagen University College. The Faculty of Health Science. Mobile +4523809597. E-mail: hlma@kp.dk

Heidi Lene Myglegård Andersen is a Danish health researcher and practitioner dedicated to developing sustainable, evidence-informed practices that strengthen collaboration across municipalities, hospitals, and social care services. Her work emphasizes participatory approaches, citizen involvement, and reducing social inequality in health. She earned her PhD from Roskilde University in 2016 with an action research–based thesis on community health strategies, revealing how top-down initiatives often miss local needs. Since then, Andersen has contributed to multiple practice-oriented research projects, bridging academic knowledge and real-world implementation while advancing critical action research to support empowerment, cross-sector learning, and contextually relevant health promotion.

ORCID: 0000-0002-7116-776X

John Andersen. Professor emitus. IMT. Roskilde Universitet
E-mail: johna@ruc.dk, https://forskning.ruc.dk/da/persons/johna

John Andersen is a professor of sociology, planning, and action research at Roskilde University. His work focuses on urban development, social inclusion, empowerment, and sustainable, democratic planning processes. Andersen is known for applying participatory action research to address challenges such as social inequality, area-based regeneration, and community engagement in deprived neighbourhoods. He has contributed extensively to research on affordable housing, age-friendly cities, and climate-sensitive urban planning. Through collaborations with municipalities, civil society, and national stakeholders, Andersen bridges academic knowledge and practical solutions. He remains a prominent voice in Danish urban planning, advocating for inclusive governance and socially just urban policies.

ORCID: 0000-0002-9762-6151

Ditte Høgsgaard Assistant Professor, Department of Regional Health, University of Southern Denmark. dmae@regionsjaelland.dk.

Ditte Høgsgaard is a Danish health researcher and practitioner known for her work in cross-sector collaboration within healthcare, particularly in developing coherent care pathways for elderly patients with multimorbidity. With a background in nursing, she is dedicated to creating sustainable, evidence-informed practices that enhance cooperation between municipalities, hospitals, and community services. Her work highlights participatory methods and strong patient involvement, ensuring that care solutions are grounded in lived experience. Høgsgaard has contributed to several practice-oriented research projects, bridging academic knowledge and practical implementation. She is recognized for promoting integrated, person-centered approaches that strengthen quality, continuity, and long-term wellbeing across healthcare sectors.

ORCID: 0000-0003-2752-881X

It takes a village: Developing action researchers through a transdisciplinary peer-learning collaborative

Keith Heggart[1], Susanne Pratt[1], Shankar Sankaran[1] and Pernille H. Christensen[1]

Abstract

This paper reports on the Participatory Action Research (PAR) Collaborative, an innovative Higher Degree Research (HDR) student development program established in 2023 at an Australian university. The initiative explored ways action research training can be integrated into HDR programs and provided a cross-faculty forum for training and peer exchange. Activities included monthly sessions, expert presentations, peer reading groups, and an online course covering action research theories, ethics, and practical challenges. This paper presents an analysis of the first two cohorts' experiences. Participants highlighted the benefits of peer learning across faculties, access to multiple informal supervisors, and increased confidence from hearing successful student stories. The findings indicate that while action research training is both feasible and valued, it requires dedicated resourcing and institutional support. Effective implementation appears to depend on cross-faculty collaboration and a diversity of student experience levels, suggesting potential for new models of HDR supervision and development across disciplines.

---

1   All authors are employed at the University of Technology Sydney

Key words: Action learning, action research, higher education, research programs

What is known about the topic?
Opportunities for Higher Degree Research (HDR) candidates to learn and practise action research (AR) or participatory action research (PAR) are limited in many university settings, despite AR's recognised value in addressing complex, real-world problems. Institutions often view AR as too contextual or time-intensive for conventional doctoral training, resulting in its marginalisation within HDR programs.
What does this paper add?
This paper presents the PAR Collaborative, an innovative cross-faculty initiative that integrates action research training into HDR programs, offering structured peer learning, access to multiple mentors, and practical engagement with AR methodologies. The analysis demonstrates that such collaborative models can build methodological confidence, foster interdisciplinary networks, and address systemic challenges in doctoral education.
Who will benefit from its content?
HDR candidates, supervisors, and university administrators seeking to diversify methodological training and support participatory research approaches will benefit from the insights and practical framework provided by the PAR Collaborative.
What is the relevance to AL and AR scholars and practitioners?
• Demonstrates a replicable model for embedding action research and peer learning in doctoral education.
• Highlights the importance of cross-disciplinary collaboration and distributed mentorship for developing AR capability.
• Offers practical strategies for overcoming isolation and methodological narrowness in HDR training
• Suggests that participatory, community-based approaches can strengthen researcher identity and prepare candidates for diverse career pathways

Received   June 2025   Reviewed   October 2025   Published   December 2025

Introduction

Despite the long-standing recognition that doctoral education should prepare researchers to address complex, real-world problems, opportunities for Higher Degree Research (HDR) candidates to learn and practise action research (AR) or participatory action research (PAR) remain limited in many university settings. While foundational contributions demonstrated the value of AR in higher education (e.g. Zuber-Skerritt & Perry, 2002), institutions have often positioned AR as too contextual, too relational, or too time-intensive to fit within conventional doctoral training (Gittens, 2019; Simonsen, 2009). As a result, students have historically been dissuaded from pursuing AR in favour of approaches viewed as more methodologically orthodox within doctoral study (Levin, 2003; Manathunga & Goozée, 2007). While recent scholarship indicates greater openness to AR, HDR candidates and supervisors still report limited structured support for building AR expertise during candidature (Alfaro-Tanco, Mediavilla & Erro-Garcés, 2023; Morales Contreras, Bellón & Barcos Redín, 2024; Pratt et al., 2024).

Paradoxically, AR continues to flourish outside the academy. In fields such as healthcare, business, and education, AR offers a rigorous, participatory, and action-oriented mode of inquiry that facilitates collaboration with diverse stakeholders (Zuber-Skerritt, 1991). Contemporary research in doctoral education reinforces the need for such relational and practice-embedded methodologies: studies show that HDR candidates benefit from developing transferable skills such as complex problem-solving, reflexivity, collaboration, and adaptability — capacities strongly aligned with AR (Brownlow, Eacersall & Martin, 2023; Skakni et al., 2025). Similarly, emerging literature highlights the value of embedded and practice-based research models, where students learn through participation in real-world, multi-stakeholder projects (Ramirez-Lovering et al., 2021), to address complex social and environmental challenges, with human and more-than-human actors. At the same time, AR is clearly positioned as a central pedagogy in some doctoral contexts (e.g., professional doctorates) and is sustained through doctoral peer collectives across disciplines (Zambo, 2011; Zhang et al., 2014). However, scholarship also suggests that access to explicit, well-supported AR pathways is uneven across doctoral training landscapes and can be constrained by institutional pressures that favour more standardised and individually oriented research trajectories (Greenwood, 2012).

This gap is increasingly difficult to justify. If HDR graduates are to contribute to diverse professional contexts, including industry and community-based roles where collaborative inquiry is expected, then universities must equip them with the methodological and relational skills required for participatory, context-responsive research. Contemporary scholarship in AR/PAR emphasises the necessity of reflexivity (Anderson & Sankofa, 2025), navigating epistemological tensions (Larrea, 2021), working across methodological traditions (Panchenko et al., 2021), and developing knowledge democratically with participants (Affouneh et al., 2023; Papadopoulou, 2021). These insights strengthen the argument that AR is not only valuable within HDR training, but also foundational to preparing researchers for the complexities of contemporary careers — academic and non-academic alike.

This paper responds to that challenge by discussing how universities might intentionally embed AR training within HDR programs. We present the Participatory Action Research Collaborative (PAR Collaborative), an innovative initiative at a large Australian university designed to help research students cultivate the skills, knowledge, and values necessary to undertake AR and PAR. The PAR Collaborative provides a year-long, cohort-based experience through which students engage with peers, academic facilitators, and practising action researchers. Its structure reflects key principles of AR, including iterative cycles of action and reflection, collective meaning-making, and sustained engagement with real-world problems.

In evaluating the PAR Collaborative, we draw on survey responses and reflective accounts from participants, alongside facilitator reflections developed through iterative cycles across two years. These data illuminate how the program supported HDR candidates to build methodological confidence, develop reflexive capacities, and form supportive peer networks — outcomes consistent with contemporary research on HDR experience (Brownlow, Eacersall & Martin, 2023) and on PAR training (Anderson & Sankofa, 2025). The analysis also enables critical consideration of how such a model may address systemic challenges in HDR training, including student isolation, supervision pressures, and the need for research approaches that bridge theory and practice.

The paper proceeds as follows. We first outline the challenges facing current HDR training and the persistent marginalisation of AR within doctoral education. We then present the design and implementation of the PAR Collaborative, detailing its structure, content, and intended outcomes. We analyse participant feedback and facilitator reflections to assess how the PAR Collaborative contributed to building AR capability and address longstanding limitations in HDR training. Finally, we discuss the implications of this model for the future of doctoral education, considering its relevance not only for improving HDR experiences but also for strengthening the methodological diversity and participatory ethos of university-based research.

Literature review

Across Australian universities, the dominant model of doctoral education has changed remarkably little over several decades. HDR students typically work within a one-to-one or one-to-two supervisory structure, where a primary supervisor assumes responsibility for the student's intellectual, methodological, and professional development. Although some institutions have introduced entrepreneurial, industry-embedded, or cohort-based programs (UTS, n.d.-a; UTS, n.d.-b), supervision continues to function largely as an apprenticeship model centred on the expertise, availability, and preferences of individual supervisors. Concerns about this model are not new; for example, Engebretson et al. (2008) argued for reconceptualising supervision as a collective and developmental relationship, yet little structural change has occurred. The longevity of the traditional model has led scholars to question its suitability for contemporary doctoral education, highlighting several significant and persistent weaknesses (McAlpine & Norton, 2006).

Methodological narrowness and the marginalisation of action research

One long-standing limitation is the methodological narrowness that can emerge when a student's orientation is shaped primarily by the dispositions and expertise of their primary supervisor. Such arrangements risk entrenching singular paradigms within supervisory lineages, making it difficult for students to pursue alternative methodologies—particularly those perceived as epistemologically or procedurally challenging. AR and PAR fall squarely into this category. Although AR has a long history and proven relevance across healthcare, education, community development, and organisational transformation (Zuber-Skerritt, 1991), supervisors unfamiliar with the approach often discourage its adoption due to concerns about rigour, ethics, and feasibility within doctoral timelines (Herr & Anderson, 2014).

Recent literature suggests that structured, program-level preparation for AR is still uneven, with scholars continuing to call for explicit AR coursework and advisor expertise to support candidates who pursue AR during candidature (Alfaro-Tanco, Mediavilla & Erro-Garcés, 2023; Morales-Contreras, Bellón & Barcos Redín, 2024; Pratt et al., 2024). This is particularly problematic given the documented epistemological challenges students face when transitioning from positivist paradigms to participatory, relational forms of inquiry. Larrea (2021), for example, demonstrates how doctoral candidates require sustained dialogue, supervision, and reflexive engagement to navigate the paradigmatic shift into AR. Without institutional support, such transitions become more difficult, leaving AR underrepresented in HDR programs despite its recognised value.

While our argument notes the limited presence of action research within HDR training, it is important to clarify that no comprehensive audit of Australian HDR programmes has been conducted to quantify the extent of this omission. Existing evidence, however, offers strong indications that AR remains marginal in doctoral education. Studies examining supervisory practices (Kiley, 2011) and reviews of doctoral program structures

(Engebretson et al., 2008; McAlpine & Norton, 2006) consistently note that practice-based and participatory methodologies receive comparatively little formal attention. These accounts suggest a wider structural pattern rather than isolated variation. In this context, the PAR Collaborative contributes to an emerging conversation about diversifying methodological training in HDR programs and responding to the unmet demand for structured support in action research. Future work could strengthen this claim further through a sector-wide audit of HDR curricula and supervisor training requirements.

Although HDR candidates benefit from broad methodological exposure, targeted methodological development also plays an important role in researcher education. AR was selected as the focus of this initiative not because it should displace other approaches, but because it represents a form of inquiry that is both widely used in professional and community settings and significantly underrepresented in HDR training. The PAR Collaborative was therefore designed not as a comprehensive methodological curriculum, but as a targeted intervention addressing a recognised gap within the broader ecology of HDR training.

Supervisory workload, capacity, and consistency

A second longstanding concern relates to supervisory workload and capacity. Supervisors are increasingly responsible for larger numbers of doctoral students and broader administrative and managerial obligations, often without proportional institutional support (Kiley, 2011; McGagh et al., 2016). This intensification can limit the time available for high-quality supervision, leading to inconsistent feedback, prolonged candidature, and emotional strain for both supervisors and candidates (Cotterall, 2013). These issues are exacerbated when students adopt unfamiliar or complex methodological approaches – such as AR – where supervisors may lack the expertise or confidence to guide an emergent, iterative process (Alfaro-Tanco, Mediavilla & Erro-Garcés, 2023). Contemporary research similarly highlights the need for more distributed supervisory models and communities of practice, where candidates can draw on multiple forms of expertise rather than relying solely on one individual (Brownlow, Eacersall & Martin, 2023; Ramirez-Lovering et al., 2021).

Isolation, wellbeing, and the HDR experience

Isolation remains one of the most consistently reported challenges of HDR study (Cotterall, 2013; Sverdlik et al., 2018). The individualised nature of doctoral work, coupled with increasingly limited supervisory availability, can undermine students' sense of belonging and identity as emerging researchers. Brownlow, Eacersall and Martin (2023), in a recent systematic review of the HDR experience in Australia, show that insufficient community engagement, inconsistent supervision, and lack of structured peer networks significantly affect student wellbeing and progression. For students pursuing methodologies that differ from disciplinary norms — such as AR — the risk of methodological and emotional isolation is heightened. These findings echo those from PAR training contexts, where scholars argue that peer learning, reflexivity, and collective inquiry form essential supports for researchers entering participatory paradigms (Anderson & Sankofa, 2025; Papadopoulou, 2021).

Employability, industry expectations, and transferable skill development

A further systemic concern relates to the mismatch between traditional doctoral training and the realities of contemporary research careers. While many candidates continue to envision academic pathways, only a small proportion ultimately remain in academia post-graduation (Cornell, 2020). Studies highlight disciplinary differences — it is often suggested that STEM candidates anticipate industry work, while social science candidates more frequently seek academic careers. Yet, across fields there is a growing expectation that HDR graduates possess complex, transferable skills such as collaboration, project management, problem-solving, and adaptive communication. These skills align closely with the competencies developed through AR/PAR, including reflexive practice, co-design, stakeholder engagement, and inquiry-in-action (Nguyen, 2025; Norton, 2018; Panchenko et al., 2021).

Skakni et al. (2025) argue that doctoral training systems must better prepare candidates for diverse career trajectories, yet the current structure of HDR programs remains dominated by academic norms such as publication, discipline-specific teaching, and traditional research methods. This emphasis marginalises relational and collaborative approaches that are increasingly valued in industry and community settings. The "crisis discourse" around employability reflects these tensions. Regardless, the shift toward diverse career outcomes underscores the need for HDR training models that cultivate the skills AR fosters—reflexivity, democratic knowledge production, inquiry into practice, and working with others to effect change.

The emerging need for cohort-based and participatory research training models

Efforts to respond to these challenges have included the introduction of multi-supervisor models, supervisory panels, and communities of practice. While promising, these arrangements often revert to the traditional model in practice due to time constraints, unclear structures, or uneven participation. More recently, research has emphasised the value of cohort-based, participatory training models that create structured opportunities for peer learning, critical reflexivity, and interdisciplinary dialogue (Anderson & Sankofa, 2025; Brownlow, Eacersall and Martin, 2023; Ramirez-Lovering et al., 2021). These innovations align closely with the principles of AR and PAR, which position learning as relational, situated, and co-constructed. Affouneh et al. (2023), for instance, illustrate how capacity-building initiatives grounded in AR can transform research cultures, disrupt epistemic hierarchies, and support more inclusive scholarly practices.

Why action research belongs in HDR training

AR offers a compelling response to the structural, methodological, and experiential limitations of current HDR models. Through iterative cycles of planning, action, reflection, and evaluation, AR integrates theory with practice, fosters reflexive researcher identities, and builds communities of inquiry. Contemporary scholarship demonstrates that AR helps doctoral candidates navigate paradigm shifts (Larrea, 2021), develop methodological competence (Norton, 2018), build relational and collaborative capabilities (Anderson & Sankofa, 2025; Papadopoulou, 2021), and undertake research that aligns with real-world complexity (Ramirez-Lovering et al., 2021). Moreover, AR is widely used outside academia, meaning HDR candidates who learn AR are better equipped for diverse professional trajectories across education, health, social services, and industry—fields that increasingly expect collaborative, applied, participatory research approaches.

For these reasons, we argue that AR training should be a core option of HDR programs at higher education institutions. The PAR Collaborative was developed to operationalise this aim by providing HDR students with an integrated, reflexive, practice-based learning environment for developing AR competencies. Below, we outline the structure and rationale of the PAR Collaborative before presenting findings from two years of participant feedback and facilitator reflection.

Innovation in researcher training: The PAR Collaborative

Rationale

The PAR Collaborative was established in 2023 at a large Australian university as an interdisciplinary initiative designed to address the well-documented structural limitations of traditional HDR training identified earlier. Its inception was grounded in a growing recognition—supported by contemporary scholarship (e.g., Anderson & Sankofa, 2025; Papadopoulou, 2021; Ramirez-Lovering et al., 2021)—that doctoral researchers require more than individualised supervision and disciplinary immersion. Instead, they benefit from structured opportunities for reflexive practice, cohort-based learning, and exposure to diverse methodological traditions, particularly those that challenge hierarchical, supervisory models of knowledge production (Affouneh et al., 2023; Larrea, 2021). The PAR Collaborative was developed as an explicit intervention to address this methodological gap while also responding to systemic challenges around isolation, supervision pressures, and career diversity.

The intention was not to promote AR as the singular or superior methodology for HDR candidates, but to create structured support for a methodological tradition that is routinely overlooked in standard research training. The PAR Collaborative complemented, rather than replaced, existing university-wide methodological offerings. By focusing on AR/PAR, the program provided depth in an area that candidates would otherwise have limited access to while remaining situated within a wider institutional landscape in which students continued to engage with a range of qualitative, quantitative, and mixed-methods approaches.

The authors of this paper—who designed and facilitated the program—came from three distinct faculties: Arts and Social Sciences, Design, Architecture and Building, and the Transdisciplinary School. This cross-faculty collaboration aligned with the institution's strategic emphasis on interdisciplinary research and its existing entrepreneurial and industry-based doctoral pathways. Bringing together facilitators from different epistemic traditions was also an intentional design choice, reflecting the need to model the pluralism and relationality central to AR. The program therefore functioned simultaneously as a training initiative and as a site of ongoing reflexive practice for its facilitators, who engaged in iterative cycles of planning, acting, reflecting, and revising the program across the 2023 and 2024 cohorts.

Design and development

The PAR Collaborative was implemented at a large Australian public university enrolling more than 40,000 students across multiple campuses, including a significant cohort of over 1,800 higher degree research candidates. As Australian universities differ substantially in size and structure from many international institutions, it is important to note that the program was situated within a large, research-intensive environment characterised by diverse disciplinary communities, a distributed supervisory workforce, and substantial numbers of international HDR candidates. These contextual features shaped both the need for and design of the PAR Collaborative. The design team comprised the authors of this paper, each bringing experience supervising HDR students who used AR, as well as experience in conducting AR projects in professional and community settings. The collaborative design approach reflected the very principles of AR itself — co-construction, reflexivity, and iterative refinement.

The program was primarily informed by established scholarship on action research training (e.g., Bradbury-Huang, 2010; Dick, 2002; Reason & Bradbury, 2008; Zuber-Skerritt & Perry, 2002), literature on doctoral education reform (McAlpine & Norton, 2006), and research on communities as mechanisms for researcher identity formation (Wenger, 1998). Several design decisions were also guided by insights from participatory research training literature in healthcare and education (Baum, MacDougall & Smith, 2006), which emphasise the need for scaffolded cycles of learning, collaboration, and supervised practice.

Three pedagogical commitments shaped the development of the curriculum:

1. **Methodological pluralism** – the goal was not to elevate AR above other methodologies but to provide HDR students with capacity in an approach that is often neglected in formal training despite its increasing use in industry and community-based research.
2. **Scaffolded cycles of learning** – monthly modules were intentionally aligned with classic AR cycles of planning, acting, observing, and reflecting (Kemmis, McTaggart & Nixon, 2014).
3. **Community-based learning** – the program privileged peer dialogue, shared inquiry, and cross-faculty learning.

The sequence of monthly topics (Table 1) emerged from mapping the core competencies required to design and implement an action research project. This mapping drew on:
- Zuber-Skerritt's (1991) action research competency frameworks,
- Reason and Bradbury's (2008) formulation of AR quality criteria, and
- professional guidelines from applied AR fields (education, health, community development).

The Canvas modules were co-designed by the author team using an iterative process across six months. Content development was guided by:
- open-access AR teaching materials (e.g., ALARA resources),
- the authors' own experience supervising AR projects.

The modules included asynchronous readings, short videos, guiding questions, and case studies sourced from contemporary AR literature.

Readings were selected according to three criteria:
- Foundational texts – to anchor students in key AR traditions and debates.
- Contemporary examples – to demonstrate current practice across sectors (education, health, industry).
- Accessibility and relevance – ensuring that readings were not overly technical and could support students with no background in AR.

Readings were reviewed annually and updated based on participant feedback, ensuring that the curriculum reflected emerging debates in the field.

The PAR Collaborative was structured around a year-long cycle, with monthly thematic modules (Table 1) and a predictable rhythm of learning activities (Figure 1).

Month	Topic
April	Getting started
May	Theoretical overview
June	Planning action research
July	Establishing and building relationships
August	Methods for data gathering
September	Methods for data analysis
October	Reflecting on cycles
November	Sharing research
December	Celebration

Table 1: The PAR Collaborative curriculum

The program's pedagogical intent extended beyond this visible curriculum. Each component was selected to foreground *participation, relationality,* and *reflexive practice* — qualities identified in recent AR/PAR literature as essential to supporting researchers transitioning into participatory inquiry paradigms (Anderson & Sankofa, 2025; Norton, 2018; Papadopoulou, 2021). For example, the reading groups were not merely discussion forums but deliberate mechanisms for cultivating a community of practice, where students could question methodological assumptions, experiment with new interpretive lenses, and learn to negotiate positionality and power within a supportive cohort. Likewise, guest lectures provided access to diverse, practice-based examples of AR, modelling the epistemic flexibility and stakeholder engagement expected of action researchers in professional contexts (Affouneh et al., 2023; Ramirez-Lovering et al., 2021).

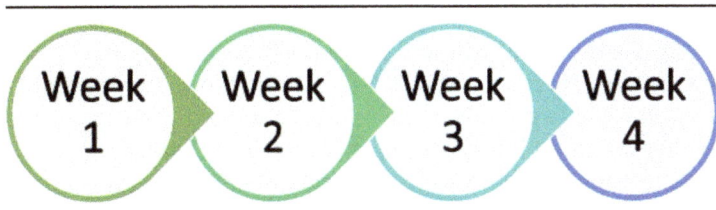

Online Learning Modules:	Peer-Based Reading Groups	Guest Lectures:	Monthly Interactive Sessions:
Self-paced courses exploring action research methodologies, ethical considerations, and case studies. The inclusion of online modules ensured that the program was accessible to all participants, regardless of their location or schedule constraints.	Small group discussions to critically engage with key literature and share diverse perspectives. These sessions also fostered a sense of shared ownership over the learning process, as participants brought their unique disciplinary perspectives to the table.	Presentations by experienced action researchers from academia and industry, as well as former HDR students who successfully employed action research in their work. Guest lectures provided participants with real-world examples and practical insights, which complemented the theoretical grounding offered in other program components.	Facilitated workshops covering theoretical foundations, ethical challenges, and practical applications of action research. These sessions were intentionally interactive to promote dialogue among participants and facilitators, ensuring that the principles of action research were modelled within the training environment.

Figure 1: Monthly cycle of the PAR Collaborative (Source: Pernille H. Christensen).

The overall rhythm of the program was intentionally cyclical, mirroring the iterative structure of AR itself. Each month culminated in a workshop that acted as a reflective "pivot point" in which students synthesised insights, articulated challenges, and planned subsequent lines of inquiry. This structure enabled students to engage with AR not only theoretically, but also *experientially* — through repeated cycles of engagement, reflection, and re-engagement. In this sense, the PAR Collaborative represented an example of what Norton (2018) describes as "PedAR": a relational pedagogy grounded in the logic of AR.

The PAR Collaborative operated as an *optional, non-award* program external to formal HDR candidature requirements at the university. Participation was voluntary and not tied to progression milestones, compulsory coursework, or supervisory sign-off. The program was offered as an enrichment opportunity designed to supplement, rather than replace, existing institutional training. Although the program received support from the relevant faculties and was advertised through central HDR communication channels, it remained an informal initiative promoted mainly through the efforts of the facilitators rather than a formally accredited component of HDR development.

This status had two implications. First, it enabled broad participation from students across the university, including those from disciplines where AR was not traditionally emphasised. Second, its optional nature meant that participants self-selected based on interest and perceived relevance, which may have contributed to high levels of motivation and engagement reported in the evaluations.

Iterative changes between Year 1 and Year 2

Consistent with the principles of action research, the PAR Collaborative evolved substantially between the 2023 and 2024 cohorts. Participant feedback, facilitator reflection, and the practical realities of delivering a year-long cross-faculty program informed targeted modifications designed to strengthen engagement, accessibility, and pedagogical coherence. These changes illustrate the PAR Collaborative's commitment to iterative refinement and highlight how participatory evaluation shaped subsequent design decisions.

Restructuring of peer learning groups

One of the clearest insights from the 2023 evaluation concerned the peer learning groups. While some students found the groups valuable, others reported inconsistent attendance, unequal participation, and difficulty sustaining momentum without formal facilitation. These challenges echoed broader debates in participatory pedagogy regarding the relationship between autonomy and scaffolding.

In response, the 2024 iteration moved away from small, fixed peer groups entirely. Instead, all participants engaged as a single large cohort during discussions, case analysis, and reflection activities. This shift ensured that students were not dependent on a small number of peers for continuity, while still providing opportunities for collaborative meaning-making. Large-group dialogue also allowed facilitators to support discussions more directly, reducing the self-efficacy burden on early-stage HDR candidates.

Change to online-only delivery

Feedback from Year 1 also highlighted the practical challenges of hybrid delivery. Although hybrid formats increased flexibility, students noted that the online and in-person experiences differed markedly. Facilitators also observed that hybrid sessions diluted opportunities for relationship-building, as online participants struggled to enter informal conversations occurring in physical rooms.

In Year 2, the PAR Collaborative shifted to an online-only delivery mode. This decision was motivated by three factors:

- ensuring equity of experience for all participants regardless of location or mode of study;
- reducing logistical complexity for facilitators; and

- strengthening the sense of a single, cohesive group by establishing a uniform modality.

The online-only format also aligned with broader institutional moves toward digital accessibility, enabling international HDR students and part-time students to participate more consistently.

Integration of Indigenous participatory methodologies

Building on feedback from 2023 and earlier facilitator reflection, the 2024 cohort included a dedicated session led by an Indigenous researcher. This session introduced yarning, relational accountability, and Indigenous participatory knowledge traditions. The inclusion of Indigenous methodologies expanded the epistemological diversity of the curriculum and signalled the need for future iterations to embed non-Western participatory methods more systematically rather than as supplementary content.

Participants and participatory dynamics

Across the two years of implementation, the PAR Collaborative remained small in scale. In 2023, ten HDR students participated; in 2024, fourteen students enrolled. Participants came from a wide range of faculties—including Arts and Social Sciences, Design, Architecture and Building, Engineering and Information Technology, and the Transdisciplinary School—and included PhD candidates as well as students enrolled in other research degrees. Participation was open, with no restrictions on mode of study (online, face-to-face, full-time, part-time), consistent with the program's commitment to accessibility and inclusivity.

A proportion of participants (n=4 in 2023) were also supervised by members of the facilitation team. This created a dual-role dynamic that required explicit attention to power, influence, and positionality. The facilitators addressed this through several practices: (1) making supervisory relationships transparent during sessions; (2) structuring activities so that supervisory and non-supervisory participants interacted evenly; (3) inviting external guest speakers to diversify voices of authority; and (4) modelling reflexivity by openly discussing methodological uncertainty, ethical dilemmas, and the limits of facilitator expertise. These approaches aligned with recommendations in the PAR literature on managing power imbalances and democratising research relationships (Anderson & Sankofa, 2025; Larrea, 2021).

Although modest in size, the diverse makeup of the cohort contributed to rich interdisciplinary dialogue and provided opportunities for students to recognise the varied forms AR can take across contexts. Participants frequently brought examples from their disciplines, workplaces, or professional histories, enabling the PAR Collaborative to function as a space where cross-sector understandings of AR could be explored and interrogated.

Data collection and analysis

To evaluate the PAR Collaborative's effectiveness and identify areas for improvement, qualitative questionnaires were administered at the end of each year. Ethics approval was sought and obtained through the university's human research ethics committee. The survey included closed and open-ended items investigating participants' perceptions of the program's value, the degree to which it supported their confidence and capability as emerging action researchers, and suggestions for refining the program design.

Five participants completed the survey in 2023 and eight in 2024. While response numbers were small, the evaluative intent was formative rather than generalisable, consistent with the iterative logic of PAR itself. Responses were analysed thematically using a collaborative coding process among the facilitation team. This allowed facilitators — who were also embedded practitioners — to engage reflexively with participant feedback, identify emergent patterns, and make corresponding adjustments to the next iteration of the program. For example, changes to the 2024 sequence were informed by 2023 feedback regarding session pacing, the level of assumed methodological expertise, and the need for clearer articulation of ethical considerations in participatory research.

Although limited, the data represent a key part of the PAR Collaborative's participatory design, with student perspectives shaping conceptual, structural, and pedagogical refinements. The analysis presented in the following section therefore reflects both participant experience and facilitator reflexivity, consistent with the collaborative, relational ethos underpinning AR/PAR.

In addition to the formal survey, each yearly cycle concluded with a dedicated reflective session in which participants collectively discussed their learning, challenges, and aspirations for the program. This final session functioned as a dialogic evaluation space aligned with PAR principles, allowing students to articulate how their understanding of AR had evolved and to identify elements of the curriculum and facilitation that supported — or hindered — their development. These discussions generated rich, practice-oriented insights that complemented the written survey data. Participants highlighted aspects of the program they found particularly valuable, such as the iterative structure, opportunities for cross-faculty dialogue, and exposure to experienced practitioners, while also offering concrete suggestions for improvement — including calls for clearer scaffolding in peer activities, more explicit examples of AR cycles in practice, and expanded attention to diverse participatory methodologies. Facilitators documented and synthesised these insights immediately following the session, and they directly informed subsequent program refinements, including shifts in modality, restructuring of collaborative activities, and enhancements to the online modules. In this way, the final reflective session formed an integral part of the program's evaluative cycle, modelling PAR's commitment to continuous, participatory learning and iterative redesign.

Participant feedback and perceived value

This section presents participant reflections on the PAR Collaborative across the 2023 and 2024 cohorts, drawing on both quantitative survey data and qualitative responses to open-ended questions. Consistent with the focus of this article, we prioritise student perspectives that speak to the pedagogical value of the PAR Collaborative and the extent to which such a model addresses systemic challenges in HDR training. While the findings reflect positive participant experiences, they also highlight important design tensions and areas for refinement, especially with regard to group processes, facilitation needs, and the realities of sustaining participatory learning environments within HDR programs.

Quantitative findings: Perceived value and impact

Participants' responses to the quantitative questionnaire (Table 2) demonstrate consistently positive perceptions of the PAR Collaborative. Across both cohorts, students strongly agreed or agreed that participation helped them to develop as researchers (12 of 13 respondents), benefitted them personally (11 of 13), and

Question	Strongly Agree	Agree	Neither Agree nor Disagree	Disagree	Strongly Disagree
Participation in the PAR Collaborative has helped you develop as a researcher	8	4	1	0	0
Participation in the PAR Collaborative has been of benefit to you personally	9	2	2	0	0
Participation in the PAR Collaborative has been of benefit to you professionally	8	4	1	0	0
I feel connected to other research students, researchers, and to the University	10	2	1	0	0
I feel integrated in a broader research culture	5	5	3	0	0

Table 2: Responses of PAR students to questions about the course and the university (n=13)

supported their professional development (12 of 13). These responses suggest that the Collaborative contributed meaningfully to the confidence, identity formation, and methodological competence of emerging action researchers.

Notably, participants also reported a heightened sense of connectedness: all respondents either agreed or strongly agreed that they felt more connected to other research students and to the university. This is significant given the persistent problem of isolation in HDR study identified in the literature (e.g., Brownlow, Eacersall & Martin, 2023; Cotterall, 2013; Sverdlik et al., 2018). Half of the respondents also reported feeling more integrated into a broader research culture, although this sentiment was less unanimous, indicating variation in how community and belonging were experienced — particularly relevant when thinking about modality and attendance patterns.

Together, these findings indicate that the PAR Collaborative met several of its intended aims: fostering a sense of community, supporting methodological development, and enhancing participant confidence. However, the quantitative data alone provide only partial insight; the following sections draw on qualitative reflections to surface deeper nuances in participant experience.

Valuable aspects of the PAR Collaborative

Belonging, community, and interdisciplinary connection

A dominant and recurring theme was the sense of belonging the PAR Collaborative cultivated. Participants emphasised the rarity of structured, cross-disciplinary spaces where HDR students could engage in sustained dialogue outside their home faculties. As one student summarised, the experience was valuable because it offered "exposure to the literature, mentoring from practitioners, [and] networking with other HDRs" (Student 1). Another noted that conversations with peers at different stages of candidature "helped contextualise what was important and what wasn't" (Student 2) and provided clarity on how action research principles could be operationalised.

This aligns with research demonstrating that communities of practice and peer learning groups can reduce isolation and support identity formation among doctoral candidates (Brownlow, Eacersall & Martin, 2023; Papadopoulou, 2021). In this case, the PAR Collaborative became a site where both methodological and affective dimensions of research could be discussed openly, with students learning not only how others were undertaking AR but also how they were making sense of themselves within its epistemological landscape.

The affective significance of belonging was particularly evident in comments emphasising emotional reassurance, confidence building, and identity transformation. One participant reflected that the PAR Collaborative "gave me confidence in my study even if I didn't quite know what I was doing," (Student 6) while another described the experience as "life-changing," explaining that it "opened my mind to the possibilities of what research CAN be, instead of what it is" (Student 5). These reflections resonate with the literature emphasising the role of reflexive, relational spaces in supporting researcher development and shift in paradigmatic stance (Anderson & Sankofa, 2025; Larrea, 2021).

Safe, supportive, and reflexive learning environments

Participants consistently highlighted the importance of the safe collaborative space created during workshops. They valued opportunities to ask questions, test ideas, challenge assumptions, and receive genuine collaborative feedback without fear of judgement. This environment appeared to be particularly significant for students navigating the uncertainties inherent in AR, where iterative cycles, emergent design, and relational ethics often depart from the certainty offered by more linear methodologies.

The cross-faculty nature of the group contributed to this ethos. Students described how interdisciplinary perspectives broadened their thinking and validated diverse research approaches. The PAR Collaborative's design thus appears to have successfully modelled the pluralistic and inclusive ethos central to AR/PAR practice.

Integration of peer learning, mentorship, and practitioner expertise

Another key strength identified by participants was the program's balanced combination of peer-based learning, facilitator guidance, and encounters with experienced action researchers. One participant described the "mix of peer learning, meeting real-life practitioners and mentors and the open discussions" (Student 11) as a particularly powerful learning experience. Others emphasised that the social bonds formed within their peer learning groups persisted beyond the formal structure of the program and directly contributed to their ongoing confidence and sense of scholarly belonging.

Less valuable aspects of the PAR Collaborative

Participant reflections also made clear that the PAR Collaborative was not universally experienced in the same way, and several areas for improvement emerged across the two cohorts. These insights are critical for refining and sustaining the program, especially given its participatory ethos.

Need for more structure and facilitation in peer learning groups

While many students valued the peer learning groups, others felt that the lack of formal facilitation affected group momentum and consistency. One participant noted that, although the self-directed format aligns with participatory principles, it presupposes high levels of "self-efficacy or motivation," which may not always be present—especially among early-stage HDR students. This tension between student autonomy and the need for scaffolding echoes broader debates in participatory pedagogy about when, where, and how facilitation supports empowerment.

Participants suggested several refinements:

- delaying the formation of groups to allow rapport to develop,
- enabling flexible group configurations rather than fixed membership, and

- ensuring academic "champions" periodically support or "wrangle" groups to maintain direction.

These suggestions reflect a nuanced understanding among participants of the balance between autonomy and support in participatory environments.

Attendance, engagement, and practical limitations

A second challenge related to attendance and the logistical realities of HDR life. Several participants remarked that irregular attendance made it difficult to sustain group discussion or maintain continuity. One participant observed that they were sometimes "left with one or no peers to meet with," (Student 12) highlighting how attrition within small groups affected the experience.

Suggestions for reducing pressure included recommending that students "self-organise a single catch-up" (Student 3) between workshops rather than maintaining formal weekly commitments. Others proposed building stronger social foundations early in the program to support sustained engagement throughout the year.

Challenges in the online environment

While some students experienced strong online connection, others did not. This reflects a wider challenge in HDR programs and online pedagogy more broadly: digital modalities can support access and flexibility but do not automatically generate authentic community. It also underscores the importance of designing hybrid and online learning environments that intentionally support relationality.

Desire for broader cross-faculty collaboration

Some participants hoped for even greater cross-faculty integration, suggesting that the PAR Collaborative could be linked to other university-wide HDR initiatives or that faculties could "advertise their forums and open them to other researchers." This highlights both the perceived value of interdisciplinary exchange and a broader institutional opportunity to leverage networks beyond the program.

Highlights and future ideas

Overall, participants described the PAR Collaborative as a deeply enriching experience that supported their development as action researchers and HDR candidates. Highlights included:

- interdisciplinary dialogue,
- exposure to practitioner expertise,
- opportunities for reflexive inquiry, and
- a sense of scholarly community rarely found in traditional supervisory structures.

Many students expressed interest in seeing the program scaled or more formally integrated into institutional HDR offerings. For example, one participant recommended a university-wide half-day HDR workshop to share insights and expand the PAR Collaborative's reach.

These forward-looking suggestions point to the potential of programs like the PAR Collaborative to contribute to systemic HDR reform — particularly in supporting methodological diversity, community building, and reflexive research practice.

Implications for future HDR training models

The purpose of establishing the PAR Collaborative was not only to provide with a structured introduction to PAR, but also to explore how programs of this kind might inform broader reforms in doctoral education. Although modest in scale, the findings point to several implications for future HDR training models, particularly in response to longstanding concerns around methodological narrowness, isolation, and preparation for diverse research careers.

Although the PAR Collaborative addressed several known challenges in HDR education, its development also highlighted the broader systemic forces shaping methodological training in universities. The marginalisation of AR in HDR programs cannot be understood solely at the program level but must be situated within the structural logics of contemporary higher education.

Performance-based research metrics, funding incentives, and institutional audit cultures tend to privilege methodologies aligned with conventional academic outputs, often disadvantaging practice-based, relational, or community-engaged approaches such as AR. Ethics approval regimes similarly reflect institutional risk aversion, creating additional hurdles for methodologies involving co-design, emergent inquiry, or ongoing negotiation with participants. These systemic pressures help explain why AR remains underutilised in Australasian HDR programs despite its relevance for real-world careers and community-engaged scholarship. The PAR Collaborative therefore operates not only as a pedagogical innovation but also as a small-scale intervention into these systems, modelling an alternative approach that foregrounds relationality, reflexivity, and collaboration in ways that push against dominant institutional norms. Future expansion of such initiatives would require stronger structural support, formal recognition within HDR policy frameworks, and ongoing dialogue about how universities can value diverse forms of research expertise and impact.

The PAR Collaborative also offers a methodological contribution to AR/AL scholarship. While AR is well established in professional and community contexts, there remain few structured models for developing AR capability within HDR programs. This initiative demonstrates how peer learning, distributed mentoring, interdisciplinary dialogue, and participatory curriculum co-design can function together as a scaffold for building AR readiness among novice researchers. In doing so, the program aligns with Action Learning, Action Research Association's principles of relational learning, reflexive practice, and collaborative inquiry, while offering a transferable framework for doctoral education. Its iterative design, annual redesign based on participant feedback, and integration of emerging methodologies — including Indigenous participatory approaches introduced in 2024 — illustrate how AR/AL pedagogy can be adapted to institutional constraints without losing its epistemic commitments. The PAR Collaborative therefore provides not only a local case but also a replicable model that may support the advancement of AR/AL training across other universities, particularly those seeking to embed participatory and socially engaged research more deeply in HDR curricula.

Peer learning as a foundation for participatory research development

A central implication concerns the role of peer learning. Participants consistently described how interactions with HDR candidates from different disciplines enhanced their understanding of AR and supported their engagement with its relational and reflexive demands. These insights reinforce existing evidence that structured peer networks can counteract the isolation characteristic of HDR study and provide a dialogic space in which candidates can negotiate methodological uncertainty. Importantly, the findings suggest that peer learning is not only socially beneficial but also foundational for researchers undertaking participatory and practice-based methodologies, which depend on shared inquiry, ongoing reflection, and collaborative sense-making.

Distributed mentorship and broadening access to expertise

The PAR Collaborative also demonstrates the value of distributed mentorship models. By engaging with facilitators, former HDR students, and experienced practitioners, participants gained access to multiple forms of expertise that extended beyond the knowledge available within a single supervisory relationship. This broadened mentorship helped students make sense of action research in practice and exposed them to a wider range of methodological and professional trajectories. Given the pressures facing supervisory staff, models that distribute mentorship across networks of practitioners, researchers, and peers may offer a viable way to extend support without increasing supervisory workload.

Supporting the development of action research capability

The findings also offer insight into how HDR candidates can develop confidence and capability in AR. Participants described gaining clarity about iterative research design, ethical considerations, and stakeholder engagement after sustained exposure to examples of AR practice. These reflections suggest that extended, cyclical, and relational forms of learning may be particularly effective for building competence in AR—far more so than short workshops or one-off seminars that present AR conceptually rather than experientially. These findings have implications for institutions seeking to diversify methodological capacity within HDR programs.

Pedagogical considerations for program design

The evaluation also highlights the design tensions that arise when implementing participatory training at scale. Some participants indicated that peer learning required more structure than initially anticipated, while online formats did not always support deep relational engagement. These insights underline the need for intentional scaffolding, including clear facilitation strategies, predictable rhythms of engagement, and attention to modality. Designing participatory spaces that can accommodate diverse levels of confidence, availability, and prior knowledge is essential if cohort-based programs are to function effectively.

Preparing researchers for diverse careers

Participants often described the program as enabling them to imagine broader research futures. This outcome aligns with calls for HDR programs to better prepare candidates for careers outside academia, where participatory, applied, and collaborative modes of inquiry are increasingly valued. By providing exposure to real-world AR practice, the PAR Collaborative helped students situate their research within wider professional contexts—an important consideration for institutions seeking to strengthen the relevance and adaptability of their HDR offerings.

Taken together, these implications suggest that participatory, relational, and community-based approaches to HDR training can contribute meaningfully to methodological development, researcher identity formation, and preparedness for diverse career pathways. They also highlight the structural and pedagogical considerations necessary for sustaining such models within contemporary university systems. Such an environment also promotes an environment for tacit knowledge exchange promoting 'Ba'[2] creating safe spaces for knowledge sharing.

Reflections as facilitators and practitioner-researchers

Our roles as designers, facilitators, supervisors, and researchers shaped the development of the PAR Collaborative and influenced how we interpreted and responded to participant feedback. Working across these overlapping positions required sustained reflexivity, particularly given the participatory and relational commitments of action research.

One insight from the first year concerned the dynamics of peer learning. We initially assumed that highly autonomous peer groups would reflect participatory principles and encourage shared ownership over the learning process. Participant feedback, however, indicated that this assumption did not hold for all students. Some groups lacked momentum or clarity, and early-stage HDR candidates often desired more structured guidance. In response, the second iteration incorporated clearer prompts, facilitated whole-cohort discussions, and flexible participation formats. These adjustments strengthened the consistency of peer engagement and demonstrated that participatory learning environments require thoughtful scaffolding rather than the withdrawal of structure.

We also reflected on the influence of supervisory relationships within the program. Because several participants were supervised by members of the facilitation team, we explicitly addressed the potential for power asymmetries by distributing facilitation roles, inviting external speakers, and making supervisor-student

---

2   In knowledge management, Ba (pronounced "bah"), introduced by Ikujiro Nonaka (see Nonaka & Konno 1998), refers to a shared space or context— physical, virtual, or mental—where individuals interact, share experiences, and create new knowledge, moving beyond mere information exchange to build shared meaning and understanding , like a team brainstorming in an office or colleagues collaborating online. It is a dynamic, emergent platform for cultivating tacit knowledge into explicit forms and back again, forming the foundation for innovation, not just a place but a dynamic relationship.

relationships transparent. While these strategies reduced some tensions, they also highlighted the persistence of hierarchical dynamics even within participatory settings. Continuing attention to positionality and influence will remain necessary in future iterations.

A further shift occurred through our expanded engagement with diverse methodological traditions. Incorporating a session led by an Indigenous researcher in the second year broadened the epistemological scope of the program and challenged us to re-examine the initial bias toward Western AR frameworks. Their contribution underscored the importance of relational accountability, yarning, and knowledge systems grounded in Country, prompting us to consider more systematic integration of Indigenous and other non-Western participatory approaches in future curriculum design.

Despite the strengths of the PAR Collaborative, our role as facilitators, supervisors, and evaluators created several tensions that merit explicit acknowledgement. AR requires ongoing reflexivity about power, positionality, and the politics of interpretation, and these dynamics were evident throughout the program. The dual role of some facilitators as both supervisors and evaluators risked influencing student participation and feedback, even with mitigating practices such as transparency and distributed facilitation. Participant reflections may therefore be partially shaped by relational obligations or perceived expectations. Declining participation across the year further highlighted the challenges of sustaining voluntary, cross-faculty initiatives alongside competing HDR demands. The evaluation design also carried limitations: the absence of baseline measures, the small sample size, and the fact that facilitators conducted the data analysis all constrain the robustness and neutrality of findings. These issues point to broader structural tensions in developing participatory researcher-development programs within university systems—particularly the reliance on goodwill, discretionary labour, and informal support rather than embedded institutional structures. A more systematic participatory evaluation process, involving students in co-analysis and co-interpretation, would strengthen future iterations and align more closely with the critical reflexivity expected in action research practice.

Overall, the program functioned as a site of professional learning for facilitators as much as for students. Iterative redesign, critical self-reflection, and engagement with participant feedback formed an ongoing cycle of inquiry into our own pedagogical practice. These experiences reaffirmed that participatory researcher development requires not only new structures for student learning, but also reflexive attention from those who design and lead such initiatives.

Beyond individual and relational outcomes, the PAR Collaborative also demonstrates potential for broader institutional and community-level impact—what is often described as third-person inquiry. While the initiative was small in scale, its cross-faculty structure provided a model for reconfiguring researcher development at a systems level by illustrating how distributed mentorship, cohort-based learning, and participatory pedagogies can complement traditional supervisory arrangements. Facilitator reflections suggest that the program encouraged supervisors to rethink aspects of their own practice, particularly in relation to reflexivity, ethical dialogue, and the relational dimensions of research training. At an institutional level, the PAR Collaborative contributed to growing interest in participatory and practice-based research approaches, signalling that such initiatives can help shift organisational cultures toward greater methodological diversity and community-engaged inquiry. Future iterations could deepen this third-person impact by connecting the PAR Collaborative with existing HDR training structures and cross-university networks, further embedding participatory principles within the broader research ecosystem.

Conclusion

This study analysed the design and early implementation of the PAR Collaborative, a year-long initiative intended to support HDR candidates in developing participatory action research capability.

The findings suggest that structured, relational, and practice-based forms of training can offer valuable support for candidates engaging with AR, particularly in navigating its methodological complexities and connecting their work to broader professional contexts. By providing access to multiple mentors, opportunities for dialogue, and iterative cycles of engagement, the PAR Collaborative offered forms of learning that extend beyond the capacities of traditional individual supervision.

The findings reaffirm the central premise captured in the paper's title: it genuinely "takes a village" to develop emerging action researchers. Participants' learning was strengthened not only through facilitator input but also through peer networks, interdisciplinary exchange, former HDR students, Indigenous researchers, and industry practitioners who contributed to the PAR Collaborative. This relational ecosystem — rather than any single instructional component — was the mechanism through which confidence, methodological capability, and research identity developed. The PAR Collaborative therefore demonstrates that cultivating an action researcher is a shared endeavour that depends on a community of practice rather than an individual supervisor, underscoring the need for HDR programs to embrace more collective, participatory modes of training.

At the same time, the evaluation revealed important considerations for future iterations. The program relied on a voluntary and self-selected cohort, and its small scale limits generalisability. The absence of baseline measures prevents conclusions about changes over time, and the bespoke questionnaire limits comparability with broader HDR research. Future work incorporating pre- and post-program data, validated survey instruments, and longitudinal tracking would enable more robust assessment of the program's outcomes and longer-term impact.

Despite these limitations, the study contributes to broader discussions about how doctoral education can evolve to better support methodological diversity, engage with participatory research traditions, and prepare candidates for varied career pathways. The findings highlight the potential benefits of cohort-based and relational approaches to methodological development, while also identifying the design and institutional challenges involved in sustaining such models. Further research across multiple institutions would help clarify how programs like the PAR Collaborative can be scaled, adapted, and integrated into wider HDR training systems.

In sum, the PAR Collaborative offers a promising model for embedding participatory principles and peer learning within doctoral education. By supporting candidates to engage with AR through extended cycles of learning, reflection, and practice, such initiatives can contribute to a richer and more responsive research training environment—one that equips emerging scholars for the epistemic, ethical, and practical demands of contemporary research.

Conflict of interest and funding statement

The authors declare that there is no conflict of interest with respect to the authorship or publication of this article. This research received no specific grant from any funding agency in the public, commercial, or not-for-profit sectors.

References

Affouneh, S., Wimpenny, K., Angelov, D., Salha, S., Khlaif, Z. N. & Yaseena, D. (2023) Fostering a culture of qualitative research and scholarly publication in a leading university in the West Bank: A Palestinian-UK capacity-building collaboration. *Higher Education Research & Development*, 42 (8), 1825-1839. https://doi.org/10.1080/07294360.2023.2209518.

Alfaro-Tanco, J. A., Mediavilla, M., & Erro-Garcés, A. (2023) Creating new knowledge while solving a relevant practical problem: Success factors for an action research-based PhD thesis in business and management. *Systemic Practice and Action Research*, 36 (6), 783–801. https://doi.org/10.1007/s11213-022-09628-6

Anderson, A. J., & Sankofa, N. (2025) Preparing for participatory action research: Professional development to support education doctoral students' critical reflexivity. *Educational Action Research*, 1-15. https://doi.org/10.1080/09650792.2025.2495684.

Baum, F., MacDougall, C., & Smith, D. (2006) Participatory action research. *Journal of Epidemiology & Community Health*, 60 (10), 854–857. https://doi.org/10.1136/jech.2004.028662.

Bradbury-Huang, H. (2010). What is good action research? *Action Research*, 8 (1), 93–109. https://doi.org/10.1177/1476750310362435.

Brownlow, C., Eacersall, D. C., & Martin, N. (2023) The higher degree research student experience in Australian universities: A systematic literature review. *Higher Education Research & Development*, 42 (7), 1608-1623.

Cornell, B. (2020). *PhD students and their careers* (HEPI Policy Note 25). Higher Education Policy Institute. https://www.hepi.ac.uk/wp-content/uploads/2020/07/HEPI-Policy-Note-25_PhD-students-careers_FINAL.pdf.

Cotterall, S. (2013) More than just a brain: Emotions and the doctoral experience. *Higher Education Research & Development*, 32 (2), 174–187.

Dick, B. (2002) Postgraduate programs using action research. *The Learning Organization*, 9 (4), 159-170.

Engebretson, K., Smith, K., McLaughlin, D., Seibold, C., Terrett, G. & Ryan, E. (2008) The changing reality of research education in Australia and implications for supervision: A review of the literature. *Teaching in Higher Education*, 13 (1), 1–15. https://doi.org/10.1080/13562510701792112 .

Gittins, C. B. (2019) Considering the future of doctoral PAR: Mapping degrees of risk, responsibility and relationships. *Educational Action Research*, 27 (5), 755–770.

Greenwood, D. J. (2012) Doing and learning action research in the neo-liberal world of contemporary higher education. *Action Research*, 10 (2), 115–132. https://doi.org/10.1177/1476750312443573.

Herr, K., & Anderson, G. L. (2014) *The action research dissertation: A guide for students and faculty* (2nd ed.). SAGE.

Kemmis, S., McTaggart, R. & Nixon, R. (2014) *The action research planner* (3rd ed.). Springer.

Kiley, M. (2011). Developments in research supervisor training. *Innovations in Education and Teaching International*, 48(4), 424–434.

Larrea, M. (2021) The PhD thesis as a threshold into action research. *Educational Action Research*, 29 (1), 5–19. https://doi.org/10.1080/09650792.2019.1702075.

Levin, M. (2003) Ph.D. programs in action research: Can they be housed in universities? *Concepts and Transformation*, 8 (3), 219–238. https://doi.org/10.1075/cat.8.3.03lev.

Manathunga, C. & Goozée, J. (2007) Challenging the dual assumption of the 'always/already' autonomous student and effective supervisor. *Teaching in Higher Education*, 12 (3), 309–322. https://doi.org/10.1080/13562510701278658.

McAlpine, L. & Norton, J. (2006) Reframing our approach to doctoral programs: an integrative framework for action and research. *Higher Education Research & Development*, 25 (1), 3–17. https://doi.org/10.1080/07294360500453012.

McGagh, J., Marsh, H., Western, M., Thomas, P., Hastings, A., Mihailova, M. & Wenham, M. (2016) *Review of Australia's Research Training System*. Report for the Australian Council of Learned Academies, www.acola.org.au.

Morales Contreras, M. F., Bellón, C. & Barcos Redín, L. (2024) Promoting insider action research: A practitioner-scholar perspective. *International Journal of Qualitative Methods*, 23, 1–18. https://doi.org/10.1177/16094069241289285

Nguyen, C. H. (2025) Teaching 'Qualitative Research Methodology' to quantitatively oriented PhD students: A practical action research study. *Vietnam Journal of Education*, 9 (Special Issue), 197–209. https://doi.org/10.52296/vje.2025.512.

Nonaka, I. & Konno, N. (1998) The concept of "Ba": Building a foundation for knowledge creation. *California Management Review*, 40 (3), 40-54. https://doi.org/10.2307/41165942.

Norton, L. (2018) *Action research in teaching and learning* (2nd ed.). Routledge. https://doi.org/10.4324/9781315147581.

Panchenko, L. F., Korzhov, H. O., Kolomiiets, T. V. & Yenin, M. N. (2021) PhD student training: Principles and implementation. *Journal of Physics: Conference Series*, 1840 (1), 012056. https://doi.org/10.1088/1742-6596/1840/1/012056.

Papadopoulou, M. (2021) A student, a practitioner or a researcher? Reconciling roles through action research. *Educational Action Research*, 29 (2), 206–225. https://doi.org/10.1080/09650792.2021.1886959.

Pratt, S., Heggart, K., Christensen, P. H., Sankaran, S. & Rees, J. (2024) Fostering participatory action research in higher degree research settings through a transdisciplinary peer-mentoring collaborative. *Systemic Practice and Action Research*, 37, 565–584. https://doi.org/10.1007/s11213-024-09691-1.

Ramirez-Lovering, D., Prescott, M. F., Josey, B., Mesgar, M., Spasojevic, D. & Wolff, E. (2021) Operationalising research: Embedded PhDs in transdisciplinary, action research projects. In R. Barnacle & D. Cuthbert (Eds.), *The PhD at the end of the world*. Springer, pp. 45–65. https://doi.org/10.1007/978-3-030-62219-0_4.

Reason, P. & Bradbury, H. (Eds.) (2008) *The SAGE handbook of action research: Participative inquiry and practice* (2nd ed.). SAGE.

Simonsen, J. (2009) The challenges for action research projects: A concern for engaged scholarship. *Scandinavian Journal of Information Systems*, 21 (1), 3–56.

Skakni, I., Kereselidze, N., Parmentier, M., Delobbe, N. & Inouye, K. (2025) PhD graduates pursuing careers beyond academia: A scoping review. *Higher Education Research & Development*. Advance online publication. https://doi.org/10.1080/07294360.2025.2515211.

Sverdlik, A., Hall, N. C., McAlpine, L. & Hubbard, K. (2018) The PhD experience. *International Journal of Doctoral Studies*, 13, 361–388.

UTS. (n.d.-a). *Entrepreneurial PhD*. University of Technology Sydney.

UTS. (n.d.-b). *Industry Doctorate Program*. University of Technology Sydney.

Wenger, E. (1998) *Communities of practice: Learning, meaning, and identity*. Cambridge University Press.

Zambo, D. (2011) Action research as signature pedagogy in an education doctorate program: The reality and hope. *Innovative Higher Education*, 36 (4), 261–271. https://doi.org/10.1007/s10755-010-9171-7.

Zhang, Z., Fyn, D., Langelotz, L., Lonngren, J., McCorquodale, L. & Nehez, J. (2014) Our way(s) to action research: Doctoral students' international and interdisciplinary collective memory work. *Action Research*, 12 (3), 293–314. https://doi.org/10.1177/1476750314534452.

Zuber-Skerritt, O. (Ed.) (1991) *Action research for change and development* (1st ed.) Routledge. https://doi.org/10.4324/9781003248491.

Zuber-Skerritt, O. & Perry, C. (2002) Action research within organisations and university thesis writing. *The Learning Organization*, 9 (4), 171–179.

Biographies

Keith Heggart

Keith is Senior Lecturer in the Faculty of Arts and Social Sciences at the University of Technology Sydney (UTS) and Director of the Centre for Research on Education in a Digital Society (CREDS). His passion lies in civics, citizenship, and digital learning, and his work spans research, teaching, and leadership in these fields. He completed his PhD at UTS in 2018 and has since coordinated and developed a range of programs, including the Master of Teaching (Secondary) and several Graduate Certificates in Learning Design and Education.

ORCID: 0000-0003-2331-1234

Susanne Pratt

Susanne Pratt is an award-winning educator, researcher and artist working at the intersection of futures-studies, ecological change, creativity and transformative learning. She is a Senior Lecturer in Transdisciplinary School at the University of Technology Sydney (UTS). Her transdisciplinary research and teaching regularly engages in action research and action learning to catalyse transformations towards regenerative futures. As the Director of Higher Degree Research (HDR) Programs at Transdisciplinary School, she is passionate about creating supportive environments to empower HDR candidates to address complex challenges through impact-oriented and engaged research.

ORCID: 0000-0002-4148-5337

Shankar Sankaran

Shankar is a Professor of Organisational Project Management at the School of the Built Environment and a Core Member of the Robotics Institute in the Faculty of Engineering and Information Technology and member of the Industry Transformation Research Stream at my School. He is also a core member of the UTS Climate Society. and Environment Research Centre (C-SERC) at my faculty. He joined UTS in 2006.

The research areas where he is creating an impact are Project Governance and Sociotechnical Systems.

His current research responsibility is as a Chief Investigator of the ARC Research Hub for Human-Robot Teaming for Sustainable and Resilient Construction. His role in this research hub is to develop sociotechnical systems principles to guide the development of human robot teaming solutions for industry.

ORCHID: [0000-0001-5485-6216](0000-0001-5485-6216)

Pernille H Christensen

Associate Professor Pernille H. Christensen received her Ph.D. in Planning, Design the Built Environment from Clemson University in South Carolina, USA. She also holds a Master of City and Regional Planning and Master of Architecture degrees from Clemson University and a Bachelor of Architecture from Mississippi State University. She has over twenty years of experience in the built environment spanning a mix of planning and design practice, financial services, and academic research experience. She has conducted research projects for industry, professional bodies, government, and quasi-government agencies, domestically and internationally. Using a multi-disciplinary approach, her research over the past decade has centred on urban sustainability and resilience, specifically focusing on strategies for improving community resilience to social and environmental disruptions and the role that the built environment plays in helping cities to mitigate impacts of these events as well as meeting targets in these areas.

ORCID: [0000-0002-1676-8570](0000-0002-1676-8570)

Enhancing EFL teacher participation in an asynchronous online forum: Integrating the Delphi technique with emancipatory participatory action research

Stuart D. Warrington[1]

Abstract

This paper reports on an emancipatory participatory action research (EPAR) project with four Japanese EFL teachers who invited me to establish, initiate, and co-moderate an asynchronous online forum to sustain teacher development disrupted by COVID-19. Despite valuing visible engagement, forum contributions remained limited. Using participatory research principles, the initiative sought to identify barriers to visible involvement and co-develop pathways toward awareness, empowerment, and renewed engagement. Innovatively, an adapted Delphi technique was integrated within EPAR to enable iterative, anonymous reflection across three cycles of participant observation. This novel synthesis uniquely addressed power imbalances as a means to foster equitable voice sharing and collective dialogue. The findings showed emergent shifts toward critical consciousness among a few participants, suggesting redistribution of communicative power enhances participation. The study offers practical support for online teacher communities and extends EPAR's application in EFL contexts

---

1   Department of British and American Studies, Nagoya University of Commerce and Business, Japan

Key words: English as a Foreign Language (EFL), Delphi technique, emancipatory participatory action research, participation, online forum

---

What is known about the topic?
Online teacher communities are valuable for professional development. However, sustaining participation and equitable engagement — especially during disruptions like COVID-19 — remains a challenge.

What does this paper add?
It introduces a novel integration of the Delphi technique within emancipatory participatory action research (EPAR). This is done to address power imbalances and promote equitable participation in online teacher forums.

Who will benefit from its content?
EFL teachers, teacher educators, and facilitators of online professional learning communities.

What is the relevance to AL and AR scholars and practitioners?
- Demonstrates how EPAR can be adapted for English language learning contexts to enhance reflective, democratic collaboration.
- Offers methodological innovation by combining EPAR with the Delphi technique for iterative, anonymous reflection.
- Provides evidence on how the redistribution of communicative power can foster engagement and critical consciousness in professional dialogue.

Received February 2022 (withdrawn) Resubmitted October 2025
Reviewed November 2025   Published December 2025

Asynchronous online forums remain widely recognized for fostering community engagement, generalized reciprocity, and cooperative social behaviours among participants (Gasell et al., 2022; Kumi, 2023; Robbins & Fairbanks, 2023). Such digital spaces provide English language educators with peer support networks to manage shared challenges in knowledge and practice (Pedraza Borbon, 2024), and, importantly, facilitate involvement among speakers for whom English is a second language (Birch & Volkov, 2007). Despite these affordances, a persistent issue is low visible participation, which is frequently under-recognized yet broadly prevalent throughout educational and professional online communities (Miyashita, 2024; Xie, Adjei & Correia, 2024).

This difficulty characterizes the asynchronous forum developed and co-moderated with four Japanese English as a Foreign Language (EFL) teachers in 2021, during the COVID-19 pandemic, when face-to-face professional development was severely limited. While participants acknowledged limited engagement as a concern, substantive contributions were minimal. Consequently, an emancipatory participatory action research (EPAR) project was undertaken with the teachers to better understand and transform partaking dynamics. To address this matter through both inquiry alongside transformation, the initiative incorporated an adapted Delphi technique within an EPAR framework. This novel integration structured collective reflection while fostering empowerment and critical re-engagement.

Guided by principles of EPAR, the study emphasized working with, for, and by participants to enable social transformation and shared knowledge construction (Jacobs, 2016; Ledwith, 2007; Reason, 2006). The Delphi adaptation contributed a structured framework for joint sense-making. This complemented EPAR's iterative cycles of reflective dialogue and shared action to create a continuous loop between research; consensus building; empowerment. Ethical safeguards, which included informed consent, confidentiality assurances, and mutual ownership of research processes, were foundational to sustaining participant trust and agency.

This paper first presents a contextual overview and literature review that foreground current investigations on asynchronous online engagement in EFL and related fields. It then details the methodology and empirical findings across three research cycles, followed by a discussion of implications for equitable online teacher communities and contributions to EPAR scholarship within language education.

Contextual background and current practice

In June 2021, four Japanese EFL teachers invited me to establish, initiate and co-moderate an asynchronous online forum using Google Groups. All members were part of the same department at a Japanese university and saw the forum as a communicative space to address two constraints. First, it aimed to compensate for the absence of regular face-to-face teacher development caused by the COVID-19 pandemic. Second, it was intended to provide flexibility in participation compared to synchronous communication platforms such as Zoom. Accordingly, I proposed the forum's purpose as "a space for sharing teaching ideas and asking and answering questions." This intent was agreed upon, and the forum was then created.

Initially, observable engagement seemed promising with 26 posts in June 2021. However, this significantly declined over subsequent months: 3 posts in July, 4 in August, 2 in September, 1 in October, 2 in November, and none in December, totalling only 38 posts in 2021 (see Table 1).

Year	J	F	M	A	M	J	J	A	S	O	N	D
2022	0											
2021						26	3	4	2	1	2	0

Table 1. Monthly number of forum postings

To understand this decline, I examined post contributions and found that two teachers made no posts at all. Of the total posts, 55% were authored by myself (including six initiating posts) and 45% by the four teachers combined (with only one initiating post). This disparity revealed my colleagues' limited visible involvement.

While participation can be interpreted differently, with many individuals engaging non-publicly as readers (Carroll & Rosson, 1996), sustaining such a forum requires a 'critical mass of contributors' (Marwell & Oliver, cited in Yeow, Johnson & Faraj, 2006, p. 968). Ridings, Gefen and Arinze (2006) likewise emphasize that a certain number of members need to contribute to maintain community life. Echoing this, my colleagues expressed concerns over their lack of visible engagement in forum-related e-mails. However, in doing so, they used collective (e.g., we) rather than individual (e.g., I / me) pronouns, which suggests they were uncomfortable acknowledging their own limited contributions.

Additionally, studies on online teacher groups since COVID-19 illustrate the growing relevance and difficulties of asynchronous forums for professional development. Abdallah and Waer (2024) discovered that social media platforms such as Facebook increasingly serve as spaces for joint continuous professional development (CPD) among EFL teachers, encouraging informal learning and the sharing of resources. However, Du et al. (2022) and Xie, Adjei and Correia (2024) report persistent participation inequalities in online discussions, often characterized by a small minority of active contributors alongside many passive observers, echoing similar concerns observed in our forum. Moreover, El-Soussi (2022) stresses how the swift transition to online teaching during the COVID-19 crisis changed teachers' professional identities and practices, warranting novel strategies to foster engagement and interaction in digital professional spaces. Such studies emphasize the call for participatory approaches and collaborative moderation techniques to sustain vibrant, empowering online teacher communities.

As a moderator, I reflected on my own posting style, which largely consisted of sharing journal articles without interactive elements such as questions or invitations to share knowledge. In retrospect, I feel this may have inadvertently discouraged my colleagues from contributing by limiting exchange. Moreover, it even resulted in a decrease in my own visible involvement from feeling a sense of isolation and frustration with a lack of response.

These observations frame the dilemma investigated in this paper, which seeks to understand and address obstacles to online forum participation through EPAR techniques, aiming to foster critical consciousness and collaborative social practices among members.

Literature review

Asynchronous forums foster EFL speakers' visible engagement by affording reflective time, diverse perspectives, and role exploration

(Bunker & Ellis, 2001; Kamhi-Stein, 2000). They also build on existing face-to-face networks to extend and amplify knowledge exchange (Fulk, Schmitz & Steinfield, 1990). Nonetheless, such established networks risk perceptions of redundancy, especially cross-culturally, where nonverbal cues and credibility assessments vary (Ibarra, 2001; Merryfield, 2001; Reeder et al., 2004).

Motivations for engaging in online communities differ, with some participants conforming to peer norms despite limited intrinsic trust (Andrews, Preece & Turoff, 2002). Common concerns include a fear of losing face, the posting of irrelevant content, or encountering cultural dissonance (Aragon, 2003; Ardichvili, Page & Wentling, 2003). The dominance of English also marginalizes less proficient speakers, limiting equal and fair engagement as well as community presence (Aragon, 2003; Yildiz & Bichelmeyer, 2003).

Together, these factors illustrate how linguistic and cultural dynamics shape engagement in EFL online spaces. To address such issues, mitigation strategies stress democratic knowledge exchange protocols and recognition of member uniqueness to bolster engagement (Ardichvili, Page & Wentling, 2003; Hew & Cheung, 2008; Krimerman, 2001; Ludford et al., 2004). Active involvement through knowledge provision, feedback, and issue-posing, sustains communities (Cross, Borgatti & Parker, 2001). Yet, intervention is often requisite for sustained participation to disrupt social inertia driven by entrenched power imbalances (McTaggart, 2007; Park, 1999). Building on these participatory principles, scholars highlight the importance of clearly defined roles and collaborative facilitation to ensure equitable involvement across research contexts.

Participation roles range from initiator to collaborator, with facilitative transition vital in participatory action research (PAR) contexts to avoid authoritarianism (de Roux, 1991; Stoecker, 1999; Stringer, 1996). This inquiry, therefore, applies such frameworks to deconstruct engagement barriers and nurture democratic, egalitarian practices.

Recent literature continues to underline the importance of social presence and active engagement for EFL asynchronous learning success (Cain, Sheehan & Taouk, 2024; Kreijns et al., 2024). The rapid shift online during COVID-19 further exposed challenges in professional development and learner interaction (Alutaybi & Alfares, 2024; Dizon & Thanyawatpokin, 2021). Within this evolving context, EPAR has also been shown to promote collaboration and learner empowerment in digital language communities (Moran et al., 2023; Pitura, 2024). Interactive instructional designs have likewise proven effective in enhancing engagement in asynchronous environments (Alshammari, 2020; Doelling & Nasrollahi, 2012).

Despite these affordances, scarce visible participation remains pervasive across educational forums (Gasell et al., 2022; Kumi, 2023; Miyashita, 2024; Robbins & Fairbanks, 2023; Xie, Adjei & Correia, 2024). Cultural and linguistic challenges further obstruct equal involvement (Ardichvili, Page & Wentling, 2003; Hew & Cheung, 2008; Aragon, 2003; Yildiz & Bichelmeyer, 2003). Preliminary studies offer some valuable insights, but pandemic-driven shifts stress the importance of integrating participatory and emancipatory approaches to sustain online EFL communities (Merryfield, 2001; McTaggart, 2007; Abdallah & Waer, 2024; Du et al., 2022; Xie, Adjei & Correia, 2024; El-Soussi, 2022; Pitura, 2024; Moran et al., 2023). Continued attention and regular updates are therefore essential to accommodate emerging linguistic, cultural, and systemic engagement factors that shape learner engagement.

Methodology

The current project aimed to answer the following questions across three cycles of EPAR, recognizing that clear questions evolve as the research progresses (Dick, 2002; Kemmis & McTaggart, 1988):

Cycle 1:

1.   Why has visible forum participation decreased?

Cycle 2:

1. What do you think of the themes posted on our forum?
2. Having acknowledged visible forum engagement as a problem, why have you rarely participated?

Cycle 3:

1. What do you think of the themes posted on our forum?
2. Building on Rampton's (1990) argument for validity of knowledge over nativeness, you seem to feel you have limited knowledge, low confidence, and English skills for visible involvement. Yet, Medgyes (2001) highlights strengths non-native EFL teachers bring. Does knowing these strengths change your feelings toward participation?
3. Why or why not?

To concretely incorporate the principles of participative research, this study was conducted collaboratively with colleagues. This was to emphasize co-identification with engagement obstacles and co-construction of solutions. By making use of iterative dialogue and consensus-building, participants served as active co-investigators shaping the inquiry process and outcomes. Such an approach aligns well with EPAR's focus on empowering stakeholders and fostering knowledge to promote social change (Reason & Bradbury, 2013).

Hence, EPAR was chosen as the framework because it works *with* participants, aiming to empower through knowledge construction and social justice (Ledwith, 2007; Reason, 1998, 2006). It explicitly acknowledges participants' perspectives and the social construction of knowledge (Carson, 1990), situating research as a tool for critical reflection and change (McTaggart, 1994, 1999). By challenging power asymmetries and fostering reflexivity, EPAR supports emancipatory outcomes despite inherent limitations and risks of reactivity (Burns, 2010; Grønhaug & Olson, 1999; Meyer et al., 1998).

Participants

Four Japanese EFL teachers from the same university and I took part. As the researcher, I assumed a dual role as participant-observer and initiator. My colleagues were fully informed about the study's aims, procedures, potential outcomes, and their rights. Moreover, each gave informed consent after receiving assurances of anonymity and the freedom to withdraw at any time without penalty (Meyer et al., cited in Vallenga et al., 2009; Park, 1999).

Ethical considerations

This inquiry was formally reviewed and subsequently approved by my university's research ethics committee. Ethical rigor was maintained via transparent informed consent procedures, assurances of participant non-recognition, and ongoing engagement. However, in spite of efforts to protect confidentiality, the small, localized participant group presented intrinsic anonymity limitations, which were clearly communicated to them.

Methods

This study utilised an adapted Delphi technique combined with participant observation to gather iterative qualitative feedback. The Delphi method's asynchronous and anonymous design helped reduce participant inhibition and social pressure, which fostered open and reflective responses and enabled the refinement of themes and interventions across multiple research cycles (Dalkey & Helmer, 1963; Landeta, 2006; Rowe & Wright, 1999). Participant observation provided further contextual insights, enriching data interpretation. Semi-structured email interviews accommodated regional dispersion and allowed timely responses, supporting sustained reflection and shared knowledge co-construction (Park, 1999; Skulmoski, Hartman & Krahn, 2007).

Data analysis

Qualitative content analysis (Elo & Kyngäs, 2007) was used to condense responses into themes, which were then member-validated (Lincoln & Guba, 1985). Iterative comparison between cycles was then undertaken to enhance their depth and trustworthiness.

Thereafter, the use of reflexive journaling (Alvesson & Sköldberg, 2010) and peer debriefing was employed to strengthen overall analytical rigor (Sandelowski, 2002). These allowed me to critically examine personal biases, contextual influences, and emergent understandings after each cycle. This sustained reflexivity fostered greater transparency and rigor in understanding participant experiences and thematic developments. Adding to this, peer debriefing sessions with my colleagues provided critical feedback on evolving interpretations, and questioned assumptions, which enriched the depth of analysis. These collaborative reflections corroborated the findings and espoused a robust understanding of the complexities common to online participatory environments.

Elaborating on these reflexive strategies, it is important to situate this approach within current methodological discussions on EPAR, especially in online and language learning contexts (El-Amin & Brion-Meisels, 2024; Upreti, Devkota & Maharjan, 2024). Scholars emphasize that EPAR entails collaboration, shared ownership, and critical reflexivity to empower participants and confront socio-political structures within educational settings (Baum, Dobrev & van Witteloostuijn, 2006; Wright, 2024). The COVID-19 pandemic hastened the reliance on online platforms for research engagement and data collection with investigators adapting protocols to maintain inclusivity and ethical rigor in remote, digital environments (Bettencourt, 2022; Wright, 2024). This warrants greater emphasis on digital literacy, participant agency, and responsiveness to socio-cultural factors influencing participation and trust in online research collaborations (Buckley-Marudas & Soltis, 2020).

Intervention

Both indirect and direct intervention strategies were employed in this project. As an indirect intervention, anonymous thematic prompts arising from member feedback were posted to stimulate critical reflection and dialogue without pressuring individuals

(Reason & Bradbury, 2013). Direct interventions involved tailored email questions informed by prior cycle findings, designed to provoke reflection on engagement barriers and encourage self-efficacy (Stringer, 2014). Interventions were discussed together, when possible, to promote shared ownership and democratic decision-making (Stoecker, 1999).

Limitations

This endeavour faced inherent constraints related to remote, asynchronous communication, where email exchanges lacked nonverbal cues and immediacy and could potentially lead to misinterpretations or delayed responses (Hine, 2000). Beyond these limitations, conducting EPAR with a small, localized group introduced further difficulties. The process of fostering critical consciousness and meaningful emancipatory change was inherently slow, nonlinear, and required sustained ethical facilitation beyond the scope of this study (Freire, 1970; Mezirow, 2000; Reason, 2006). Although safeguards such as informed consent and confidentiality were rigorously applied to all participants, the small group size limited the full guarantee of anonymity and may have hindered their openness and empowerment (Minkler & Wallerstein, 2011).

Moreover, enduring power asymmetries remain difficult to eliminate in contexts where the researcher has multiple roles as these influence participant dynamics (Canagarajah, 2006). External cultural, institutional, and linguistic barriers further limit the extent that emancipation is readily achievable (Kemmis, 2007). Therefore, rather than representing a complete realization of emancipatory aims, the shifts observed here should be thought of as important initial steps. Future cycles of participatory research and intervention will be essential to deepen empowerment and promote wider systemic transformation.

Findings

This section presents the findings from three action research cycles exploring teacher engagement in an asynchronous online forum.

Each cycle provided evolving insights into barriers and possibilities for participation.

Cycle 1 outcomes

The first cycle explored my colleagues' perceptions regarding their limited visible involvement in the asynchronous online forum. Four interrelated themes emerged: perceived knowledge and confidence deficits, heavy workload, limited English proficiency, and personal choice.

Participants voiced self-doubt and indecision around contributing. This is in line with current research illustrating teacher uncertainty and reluctance to post because of a perceived lack of expertise or social anxiety in digital environments (Xie, 2023; Goudarzi et al., 2023). Indeed, one colleague stated that, "Some members aren't confident about their knowledge... they're not comfortable sharing anything," exemplifying typical self-marginalization.

Workload stresses were also evident, supporting findings that competing professional duties greatly impede asynchronous dynamics in online teacher communities (Chen & Zhao, 2023; Liu, Sun & Cui, 2025). Another colleague stated, "Because we are busy doing work," denoting the presence of ongoing practical concerns.

Language proficiency further remains a crucial barrier. One colleague said, "Some of us struggle to understand what's posted, so we just don't respond," substantiating studies on the limiting impact of English proficiency in EFL asynchronous forums and the need for more tailored facilitation to lessen linguistic obstacles (Al-Khresheh, 2025; Li, Majumdar & Ogata, 2025).

While the group acknowledged the trouble with engagement, use of communal terms such as "our group" or "we" rather than personal ownership suggested limited internalization of responsibility — reflecting broader difficulties in cultivating individual agency and critical consciousness in online communities (Jonsson, 2023; Kornienko et al., 2024).

Between January 2 and 8, 2022, the anonymized themes were posted on the forum as an indirect intervention with automatic email notifications. However, from observations, visible engagement did not increase, underlining the complexity of behaviour change in asynchronous spaces (Xie, Adjei & Correia 2024; Zhang & Liu, 2023). Building on these initial insights into practical and linguistic barriers, Cycle 2 delved deeper into the discreet emotional and motivational factors shaping engagement

Cycle 2 outcomes

Cycle 2 explored invisible, emotional and motivational barriers. Deeper reflections surfaced related to inconspicuous barriers to participation such as emotional and motivational factors. Colleagues expressed cautious optimism and tentative steps toward re-engagement despite the presence of apprehensions.

One participant noted, "I am trying to read more posts but still unsure about my contributions," illustrating persistent self-efficacy dilemmas.

Recent studies point to the strong significance of emotional engagement and peer support in sustaining asynchronous involvement in EFL forums (Liu & Zhou, 2024; Zhong, Ismail & Lin, 2025). Teacher emotional support has been found to encourage enjoyment and motivation, while peer networks heighten learner confidence and decrease feelings of isolation (Zhang & Hu, 2025).

Participants also mentioned a need for more explicit forum rules and more structured facilitation. These parallel documented effective practices emphasizing agenda-setting and supportive interaction (Voltage Control, 2024; Wang & Zhang, 2023).

An enhanced sense of social presence was also prominent, which reflects an increasing awareness that community links and relational dynamics affect partaking choices (Hoch & Fry, 2024; Kreijns et al., 2024). However, the gap between valuing involvement and actual posting persisted, remaining a familiar challenge in asynchronous online learning environments (Cain, Sheehan & Taouk, 2024).

Throughout the iterative cycles, participants poignantly reflected on their evolving understanding of engagement barriers. For example, colleague D shared, "Initially, I felt hesitant to post because my English felt inadequate, but through the cycles, I realized my experiences hold value beyond language proficiency." Such reflections reveal how the adapted Delphi technique not only surfaced practical constraints but also fostered gradual critical consciousness and self-efficacy.

Another relayed a moment of realization during Cycle 2 interventions: "Recognizing that sharing responsibility eased my anxiety was transformative and motivated me to engage more openly." These narratives convey the deeply personal and emergent nature of engagement in online forums.

The following table summarizes key colleague statements reflecting nascent shifts toward critical consciousness during the second cycle.

Colleague B	Colleague D
"I think saying I'm busy with work is sometimes an excuse. Maybe I just don't realize I learn best through questions and engaging in discussion."	"I think this group shouldn't just benefit me. It's embarrassing knowing my colleagues give me useful information and I cannot do the same."

Table 2. Signs of movement toward critical consciousness

These reflections highlight participants' emerging awareness of their dual role as learners and contributors. To add, they mark their initial steps towards critical consciousness and shared responsibility within the forum. This aligns well with EPAR principles emphasizing individual and collective ownership as foundational to sustainable change (Herr & Anderson, 2015; Reason, 2006). Consequently, with a growing awareness of emotional and social barriers, Cycle 3 focused on emerging shifts toward collective responsibility and co-moderation strategies to sustain participation.

Cycle 3 outcomes

Cycle 3 revealed a promising shift toward collective commitment and co-moderation as strategies to visible engagement. Colleagues demonstrated a greater willingness to share responsibility for sustaining forum activities, moving from individual hesitation to a communal sense of ownership. For example, one stated, "If we share the task, I feel more comfortable participating," reflecting improved agency and collaborative spirit.

This aligns with recent studies showing that distributed leadership and shared facilitation sustain participation, boost inclusivity, and reduce moderator fatigue in online learning communities (Wang & Zhang, 2023; Zhao, Li & Kang, 2025). Such shared ownership embodies EPAR principles of participant co-creation and shared change agency (Herr & Anderson, 2015; Reason, 2006).

Nevertheless, persistent obstacles related to workload and English proficiency constrained full engagement. This is consistent with common predicaments in asynchronous EFL forums that call for context-sensitive support mechanisms (Li, Majumdar and Ogata, 2025; Tawalbeh & Al-Husban, 2023).

Table 3 illustrates the direct intervention's impact on participant attitudes, showcasing critical self-reflection and a move toward active engagement intentions. These reflections represent important steps toward overcoming earlier barriers.

Colleague B	Colleague D
"I should communicate more as being part of a group is not only receiving but also giving."	"They slightly change my feelings. I now see I can bring new knowledge to our forum and be a giver, not just a receiver."
"I have often used my English as a reason not to participate, but I think this is an excuse."	"I realise now that I don't participate very much and should contribute more."

Table 3: Impact of direct intervention

Reflecting on the indirect intervention's limited visible success highlights the intricate, multifaceted dynamics influencing asynchronous engagement. These insights signify the importance of ethical facilitation, trust-building, and participant-led approaches in fostering sustainable involvement (Cain, Sheehan & Taouk, 2024; Merryfield, 2001).

In sum, Cycle 3 reveals an evolving ethos that balances emerging collective agency with practical constraints. This indicates a need for continuous, participatory facilitation efforts that empower participants while addressing linguistic and structural barriers. In addition, the iterative cycles collectively reveal a developmental trajectory where participants move from perceived external barriers toward internalized agency and communal ownership. This progression highlights the transformative potential of ethical, participant-driven facilitation in overcoming linguistic, workload, and motivational challenges inherent in asynchronous online learning environments. Future studies should explore tailored facilitation strategies that address linguistic and contextual factors, while ongoing interventions should prioritize fostering collective responsibility and active engagement.

Discussion

Cultural adaptations of EPAR

This project's findings are contextualized within broader socio-cultural and institutional frameworks affecting participant engagement. Japan's cultural paradigms—group harmony (wa), indirect communication, and hierarchical respect—required adaptation of EPAR methods to keep relational harmony while encouraging critical reflection (Mori, 2019; Richardson et al., 2023). These cultural adjustments appear to have contributed to nuanced critical consciousness among participants.

EPAR, predominantly developed within Western contexts emphasizing direct communication and explicit agency (Herr & Anderson, 2015), is enriched here by considering culturally distinctive Japanese values such as group harmony and indirect communication. By acknowledging these non-Western ways of understanding the world, EPAR can draw upon a wider set of methods that illuminated different perspectives on knowledge and change (Richardson et al., 2023). These cultural influences also interact with systemic barriers, showing how involvement is shaped not just between people, but within organizations as well.

Systemic and pandemic contexts

The findings also have to be understood against a backdrop of broader educational and cultural system, where institutional policies, technological infrastructure, and societal norms jointly influence teacher engagement in online forums. The impact of COVID-19 restrictions reshaped professional development modes, while hierarchical norms within Japanese educational institutions affected interaction dynamics. A systemic lens (Checkland, 1999) reveals that sustained change requires intervention beyond individual agency to include organizational support and cultural adaptation.

The study further resonates with the cyclical model of action learning, showing how iterative reflection and shared problem-solving promote double-loop learning (Revans, 1982). By integrating EPAR principles, the initiative allowed for transformative praxis that interrogated power relations influencing participation, thereby balancing micro-level empowerment and macro-level systemic awareness (Reason & Bradbury, 2013). The cultural adaptations evidenced here support the need for context-sensitive facilitation, echoing Earley and Ang's (2003) argument that cross-cultural effectiveness requires critical reflexivity in cross-cultural practice.

Indeed, teacher engagement concerns exacerbated by systemic pandemic stresses—organizational disruptions, workload surges, rapid technology adoption, and professional identity tensions—indicate that sustainable contribution interventions need to extend beyond motivation to include institutional policy and resource allocation (Tsubono & Mitoku, 2023).

Methodological advances

The fusion of EPAR with an adapted Delphi technique advances the former by addressing power asymmetries and promoting equitable voice in digital settings (Kemmis, McTaggart & Nixon, 2014; Reason & Bradbury, 2013). The iterative, anonymous reflection enabled through the integration of the Delphi method supports praxis addressing dominance hierarchies and social desirability bias (Sattler et al., 2022). This approach not only strengthens EPAR's collaborative ethos but also extends its transformative impact across individual, communal, and systemic levels (Greenwood & Levin, 2007).

In this study, the emergent critical consciousness shift among two participants and recognition of institutional barriers—workload, policy, and resource constraints—illustrates the complex multifaceted nature of emancipatory change in professional online communities. Integrating cultural sensitivity with institutional responsiveness can broaden EPAR's transformational potential in shaping educational policy and practice.

While this study focused on a small, localized group of Japanese EFL teachers, the issues and emergent strategies have conceivably wider implications for many asynchronous online communities and EPAR projects worldwide. The practical integration of the Delphi technique with EPAR offers a reproducible methodology for fostering critical consciousness and collaborative ownership across diverse cultural and linguistic settings. These findings encourage long-term investigations to refine participatory interventions while promoting sustainable online communities and expanding emancipatory outcomes.

Practical interventions

Building on participants' lived experiences, this study translates their insights into concrete practice improvements through tailored interventions incorporating anonymity and iterative feedback. These measures alleviated anxieties around visible involvement, fostering a more supportive environment that empowered hesitant members. The need for shared ownership can therefore inform the development of distributed leadership and peer mentorship models, which sustain community vitality beyond individual facilitation. Shifting from a single moderator to shared ownership is also key to making online teacher communities more inclusive and keeping members engaged. By opening up involvement in this way, teachers can build knowledge together — the very foundation of a thriving online community (Garrison, 2017; Hess & Saxberg, 2021).

Anonymity and ongoing feedback are also important as they help to ease participants' anxiety and encourage them to engage. At the same time, spreading leadership responsibilities among members allows them to share roles and hold joint ownership which can lighten the workload for moderators while strengthening community resilience. Finally, tailored interventions targeting language confidence and intercultural communication needs can address essential conditions for sustained involvement.

Future directions & conclusion

Future cycles will aim to deepen participant engagement and sustain transformative learning processes. It is hoped that these cycles will generate deeper, long-term insights into how empowerment lasts and systemic change takes root. Ongoing support therefore remains essential for our forum, inclusive of language development, confidence-building, managing workload pressures, promoting peer mentorship, and sharing leadership to help our community grow (Wang & Zhang, 2023; Zhao, Li & Kang, 2025).

To inform replication in other contexts, the culturally responsive features developed here (i.e., anonymity, iterative feedback, and distributed facilitation) can serve as flexible process guidelines for multilingual and multicultural professional settings. In addition, these same principles may translate into broader strategies fostering psychological safety and equitable voice in culturally and linguistically diverse settings. Cycles hereafter could expand to larger cohorts or regional professional learning networks. Such extensions could examine how these adaptations shape reflective dialogue and power redistribution across cultural as well as linguistic boundaries.

Additional studies will be needed to build on these directions by examining longitudinal effects, identifying best practices, and optimizing technology-enhanced facilitation across diverse cultural and linguistic contexts. Cross-site comparisons may further clarify how EPAR's emancipatory potential can inform sustainable, multilingual knowledge communities.

In summary, this initiative illustrates that keeping EFL teachers engaged in asynchronous forums calls for thoughtful, ethical, and multifaceted approaches that build critical awareness and take practical concerns into account. With shared ownership and participatory facilitation, these forums can develop into vibrant communities where every voice is heard and valued. By framing these adaptable principles as scaffolding for equitable participation, EPAR can contribute to a globally attuned praxis that nurtures inclusion, reflection, and shared agency.

Critical reflection

This EPAR initiative delved into the underlying causes of limited visible participation among four Japanese EFL teachers in an asynchronous online forum designed to support professional development during the COVID-19 crisis. Three interwoven reasons consistently surfaced: personal choice, perceptions of insufficient knowledge and confidence, and perceived limited English proficiency. These factors formed a complex web sustaining engagement hesitancy.

Of the four participants, two colleagues, B and D, showed promising signs of developing critical consciousness, growing in awareness of their role and agency in the community. Their reflections suggest that interventions engendered shifts in their mindsets and prompted reconsideration of contribution norms. However, their contributory voices remain largely muted alongside colleagues A and C, highlighting that critical consciousness requires nurturing within supportive structures (Reason, 2006).

The ongoing lack of visible engagement underscores how transformative change in education is often slow, nonlinear and sometimes uncertain (Mezirow, 2000). Developing critical awareness takes time, along with consistent and ethical facilitation that fosters trust and empowers participants (Block, 2018; Freire, 1970).

Hence, this study's findings raise important questions about power relations and community ownership in professional online spaces. Moderating an online forum is a task that cannot rest solely upon one person, but rather it calls for shared communal responsibility. To advance inclusivity and sustainability, the fair redistribution of power and participation roles among members is key. Such democratization underpins the co-construction of a knowledge community in which responsibility, voice and vitality are shared, which, in turn, strengthens both sustainability and professional growth (Garrison, 2017; Hess & Saxberg, 2021).

Funding

This research received no external funding or grant.

References

Abdallah, M.M.S. & Waer, H. (2024) Exploring the informal online practices of in-service English Language Teachers on Facebook as part of their continuing professional development. *CDELT Occasional Papers in the Development of English Education*, 87 (1). Available at https://journals.ekb.eg/article_384382.html.

Al-Khresheh, M.A. (2025) The subtle power of nonverbal communication in EFL classrooms: An observational study on Jordanian students. *Australian Journal of Applied Linguistics*, 8 (5), 123–140. Available at https://doi.org/10.29140/ajal.v8n5.102643.

Alshammari, S.R. (2020) "Writing to learn or learning to write": A critical review of EFL writing practices in Saudi universities. *Research in Education and Learning Innovation Archives*, 24, 1–22. Available at https://doi.org/10.7203/realia.24.15867.

Alutaybi, M.M. & Alfares, N.S. (2023) A prospective study for exploring Saudi EFL learners' strategic listening skills through Netflix from teachers' perspectives. *Arab World English Journal*, 14 (4), 300–311. Available at https://dx.doi.org/10.24093/awej/vol14no4.18.

Alvesson, M. & Sköldberg, K. (2010) *Reflexive methodology: New vistas for qualitative research*. 2nd ed. London; Sage.

Andrews, D., Preece, J. & Turoff, M. (2002) A conceptual framework for demographic groups resistant to online community interaction. *International Journal of Electronic Commerce*, 6 (3), 9–24. Available at https://www.dhi.ac.uk/san/waysofbeing/data/communities-murphy-andrews-2002.pdf.

Aragon, S.R. (2003) Creating social presence in online environments. In Aragon, S.R. (Ed.) *Facilitating learning in online environments*. Jossey-Bass, pp. 57–68. Available at: https://doi.org/10.1002/ace.119.

Ardichvili, A., Page, V. & Wentling, T. (2003) Motivation and barriers to participation in virtual knowledge-sharing communities of practice. *Journal of Knowledge Management*, 7(1), 64–77. Available at https://doi.org/10.1108/13673270310463626.

Baum, J.A.C., Dobrev, S.D. & van Witteloostuijn, A. (Eds.) (2006) *Advances in strategic management: Vol. 23: Strategy and ecology*. Elsevier.

Bettencourt, L.M.A. (2022) The digital transformation of metropolises. *Metropolis Observatory Issue Paper*, pp. 3–14. Available at https://www.metropolis.org/wp-content/uploads/01_Observatory_Issue-Paper-08_Digital-transformation-metropolises.pdf.

Birch, D. & Volkov, M. (2007) Assessment of online reflections: Engaging English second language (ESL) students. *Australasian Journal of Educational Technology*, 23 (3), 291–306. Available at https://doi.org/10.14742/ajet.1254.

Block, D. (2018) The political economy of language education research (or the lack thereof): Nancy Fraser and the case of translanguaging. *Critical Inquiry in Language Studies*, 15 (4), 237–257. Available at https://doi.org/10.1080/15427587.2018.1466300.

Buckley-Marudas, M. F. & Soltis, S. (2020) What youth care about: Exploring topic identification for youth-led research in school. *The Urban Review*, 52 (2), 331–350. Available at https://doi.org/10.1007/s11256-019-00530-5.

Bunker, A. & Ellis, R. (2001) Using bulletin boards for learning: What do staff and students need to know in order to use boards effectively? In Hermann, A. & Kulski, M.M. (Eds.) *Expanded horizons in teaching and learning. Proceedings of the 10th Annual Teaching Learning Forum*, Curtin

University of Technology, Perth, 7-9 February 2001. Available at https://ro.ecu.edu.au/ecuworks/4897/.

Burns, A. (2010) Action research. In Paltridge, B. & Phakiti, A. (Eds.) *Continuum companion to research methods in applied linguistics*. Continuum International Publishing Group, pp. 80–97.

Cain, M., Sheehan, H. & Taouk, S. (2024) Preservice teachers' experiences of relationality in asynchronous online learning. *Journal of Further and Higher Education*, 48 (3), 449–463. Available at https://doi.org/10.1080/1359866X.2025.2539248.

Canagarajah, S. (2006) *Ethnographic methods in language policy*. Routledge.

Carroll, J.M. & Rosson, M.B. (1996) Developing the Blacksburg electronic village. *Communications of the ACM*, 39 (12), 69–74. Available at https://doi.org/10.1145/240483.240498.

Carson, T. (1990) What kind of knowing is critical to action research? *Theory Into Practice*, 29 (3), 167–173. Available at https://doi.org/10.1080/00405849009543450.

Checkland, P.B. (1999) *Systems thinking, systems practice*. John Wiley & Sons Ltd.

Chen, B. & Zhao, C. (2023) More is less: Homeroom teachers' administrative duties and students' achievements in China. *Teaching and Teacher Education*, 119, 103982. Available at https://doi.org/10.1016/j.tate.2022.103857.

Cross, R., Borgatti, S.P. & Parker, A. (2001) Beyond answers: Dimensions of the advice network. *Social Networks*, 23 (3), 215–235. Available at https://doi.org/10.1016/S0378-8733(01)00041-7.

Dalkey, N.C. & Helmer, O. (1963) An experimental application of the Delphi method to the use of experts. *Management Science*, 9 (3), 458–467. Available at https://doi.org/10.1287/mnsc.9.3.458.

de Roux, G.I. (1991) Together against the computer: PAR and the struggle of Afro-Colombians for public service. In Fals Borda, O. & Rahman, M.A. (Eds.) *Action and knowledge: Breaking the monopoly with participatory action-research*. Apex Press, pp. 37–53. Available at https://ferlernen.wordpress.com/wp-content/uploads/2015/03/fals-borda-action_and-knowledge.pdf.

Dick, B. (2002) Postgraduate programs using action research. *The Learning Organization*, 9 (4), 159–170. Available at https://doi.org/10.1108/09696470210428886.

Dizon, G. & Thanyawatpokin, B. (2021) Emergency remote language learning: Student perspectives of L2 learning during the COVID-19 pandemic. *The JALT CALL Journal*, 17 (3), 349–370. Available at https://doi.org/10.29140/jaltcall.v17n3.431.

Doelling, M. & Nasrollahi, F. (2012) Building performance modeling in non-simplified architectural design: Procedural and cognitive challenges in education. In *Proceedings of the 30th eCAADe Conference*, Leuven; KU Leuven, pp. 116-124. Available at: https://papers.cumincad.org/cgi-bin/works/paper/ecaade2012_116.

Du, Z., Wang, F., Wang, S. & Xiao, X. (2022) Enhancing learner participation in online discussion forums in massive open online courses: the role of mandatory participation. *Frontiers in Psychology*, 13, 819640. Available at https://doi.org/10.3389/fpsyg.2022.819640.

Earley, P.C. & Ang, S. (2003) *Cultural Intelligence: Individual interactions across cultures*. Stanford University Press.

El-Amin, A. & Brion-Meisels, G. (2024) Editorial: Emancipatory inquiry in educational research: models and methods for transformational learning. *Frontiers in Education*, 9, 1396662. Available at https://www.frontiersin.org/journals/education/articles/10.3389/feduc.2024.1396662/full.

Elo, S. & Kyngäs, H. (2007) The qualitative content analysis process. *Journal of Advanced Nursing*, 62 (1), 107–115. Available at https://doi.org/10.1111/j.1365-2648.2007.04569.x.

El-Soussi, A. (2022) The shift from face-to-face to online teaching due to COVID-19: Its impact on higher education faculty's professional identity. *International Journal of Educational Research Open*, 3, 100139. Available at https://doi.org/10.1016/j.ijedro.2022.100139.

Freire, P. (1970) *Pedagogy of the oppressed*. Penguin Books.

Fulk, J., Schmitz, J. & Steinfield, C. (1990) A social influence model of technology use. In Fulk, J. & Steinfield, C. (Eds.) *Organizations and communication technology*. Sage, pp. 117–142. Available at: https://doi.org/10.4135/9781483325385.n6.

Garrison, D.R. (2017) *E-learning in the 21st century: A community of inquiry framework for research and practice*. 3rd ed. Routledge.

Gasell, C., Lowenthal, P.R., Uribe-Flórez, L.J. & Ching, Y.H., 2022 Interaction in asynchronous discussion boards: a campus-wide analysis to better understand regular and substantive interaction. *Education and Information Technologies*, 27 (3), 3421-3445. Available at https://doi.org/10.1007/s10639-021-10745-3.

Goudarzi, E., Hasanvand, S., Raoufi, S. & Amini, M. (2023) Teachers' experiences of teaching during the COVID-19 pandemic. *PLoS ONE*, 18 (11), e0287520. Available at https://doi.org/10.1371/journal.pone.0287520.

Greenwood, D.J. and Levin, M. (2007) *Introduction to action research: Social research for social change*. 2nd ed. Sage Publications.

Grønhaug, K. and Olson, O. (1999) Action research and knowledge creation: merits and challenges. *Qualitative Market Research: An International Journal*, 2 (1), 6–14. Available at https://doi.org/10.1108/13522759910251891.

Herr, K. & Anderson, G.L. (2015) *The action research dissertation: A guide for students and faculty*. Sage.

Hess, A. & Saxberg, S. (2021) *Presence in online learning: How to design and facilitate impactful digital learning experiences*. Routledge.

Hew, K. F. & Cheung, W. S. (2008) Attracting student participation in asynchronous online discussions: A case study of peer facilitation. *Computers & Education*, 51 (3), 1111–1124. Available at https://doi.org/10.1016/j.compedu.2007.11.002.

Hine, C. (2000) *Virtual ethnography*. Sage.

Hoch, M.L. & Fry, M. (2024) Establishing social presence through online interactions: A case study in a literacy clinic. *i.e.: Inquiry in Education*, 16 (1), Article 4. Available at https://digitalcommons.nl.edu/ie/vol16/iss1/4.

Ibarra, R. A. (2001) *Beyond affirmative action: Reframing the context of higher education*. University of Wisconsin Press.

Jacobs, S. (2016) The use of participatory action research within education—Benefits to stakeholders. *World Journal of Education*, 6 (3), 48-55. Available at https://doi.org/10.5430/wje.v6n3p48.

Jonsson, J. (2023) Exploring the social and spatial role of social media for community entrepreneurship: A rural grocery store case study. *Entrepreneurship & Regional Development*, 36 (7-8), 1054–1070. Available at https://doi.org/10.1080/08985626.2023.2287696.

Kamhi-Stein, L. D. (2000) Looking to the future of TESOL teacher education: Web-based bulletin board discussions in a methods course. *TESOL Quarterly*, 34 (3), 423–455. Available at https://doi.org/10.2307/3587738.

Kemmis, S. (2007) Critical theory and participatory action research. In Reason, P. and Bradbury, H. (Eds.) *The SAGE handbook of action research: Participative inquiry and practice. 2nd ed. SAGE, pp. 121-138. Available at https://researchoutput.csu.edu.au/ws/portalfiles/portal/9916955/CSU268100.pdf.

Kemmis, S. & McTaggart, R. (1988) *The action research planner*. Deakin University.

Kemmis, S., McTaggart, R. & Nixon, R. (2014) *The action research planner: Doing critical participatory action research*. 3rd ed. Springer.

Kornienko, O., Burson, E., Young, D. L., Godfrey, E. & Garner, P. (2024) Conscious selection: Critical consciousness informs cultivation of social networks in a living and learning community. *Social Development*, 34 (1), e12780. Available at https://doi.org/10.1111/sode.12780.

Kreijns, K., Yau, J., Weidlich, J. & Weinberger, A. (2024) Toward a comprehensive framework of social presence for online, hybrid, and blended learning. *Frontiers in Education, Section Digital Education*, 8. Available at https://doi.org/10.3389/feduc.2023.1286594.

Krimerman, L. (2001) Participatory action research: Should social inquiry be conducted democratically? *Philosophy of the Social Sciences*, 31 (1), 60–83. Available at https://doi.org/10.1177/004839310103100104.

Kumi, R. (2023) Discussion forums and student engagement: A social presence perspective. *Distance Education Society Journal*, 12 (4), 56-67. Available at https://doi.org/10.1111/dsji.12299.

Landeta, J. (2006) Current validity of the Delphi method in social sciences. *Technological Forecasting and Social Change*, 73 (5), 467–482. Available at https://doi.org/10.1016/j.techfore.2005.09.002.

Ledwith, M. (2007) On being critical: Uniting theory and practice through emancipatory action research. *Educational Action Research*, 15 (4), 597–611. Available at https://doi.org/10.1080/09650790701664021.

Li, H., Majumdar, R. & Ogata, H. (2025) Self-directed extensive reading with social support: effect on reading and learning performance of high and low English proficiency students. *Research and Practice in Technology Enhanced Learning*, 20 (25). Available at https://pdfs.semanticscholar.org/800b/296831d9cea29bef823d944d98de8c02f58f.pdf.

Lincoln, Y. S. & Guba, E. (1985) *Naturalistic enquiry*. Sage.

Liu, D., Sun, Z. & Cui, Y. (2025) How institutional support enhances teacher engagement in online teaching: chain mediation effects of digital self-efficacy and negative emotions. *Frontiers in Psychology*, 16, 1601764. Available at https://doi.org/10.3389/fpsyg.2025.1601764.

Liu, Q. & Zhou, W. (2024) The impact of teachers' emotional support on EFL learners' online learning engagement: The role of enjoyment and boredom. *Acta Psychologica*, 250, 104504. Available at https://www.sciencedirect.com/science/article/pii/S0001691824003822?via%3Dihub.

Ludford, P.J., Cosley, D., Frankowski, D. & Terveen, L. (2004) Think different: Increasing online community participation using uniqueness and group dissimilarity. In *Proceedings of the SIGCHI Conference on Human Factors in Computing Systems* (CHI '04), pp. 631–638. Available at https://doi.org/10.1145/985692.985772.

McTaggart, R. (1994) Participatory action research: Issues in theory and practice. *Educational Action Research*, 2 (3), 313–337. Available at https://doi.org/10.1080/0965079940020302.

McTaggart, R. (1999) Reflection on the purposes of research, action, and scholarship: A case of cross-cultural participatory action research. *Systematic Practice and Action Research*, 12 (5), 493–511. Available at https://doi.org/10.1023/A:1022417623393.

McTaggart, R. (2007) The role of professors in participatory action research. In Santos Caicedo, D. A. & Todhunter, M. (Eds.) *Festschrift for Orlando Fals Borda*, Universidad de La Salle Press, pp. 1–12.

Medgyes, P. (2001) When the teacher is a non-native speaker. In Celece-Murcia, M. (Ed.) *Teaching English as a second or foreign language*. Heinle & Heinle, pp. 429–442.

Merryfield, M. (2001) The paradoxes of teaching a multicultural education course online. *Journal of Teacher Education*, 25 (4), 283–299. Available at https://doi.org/10.1177/0022487101052004003.

Mezirow, J. (2000) *Learning as transformation: Critical perspectives on a theory in progress*. Jossey-Bass.

Meyer, L., Park, H. S., Grenot-Scheyer, M., Schwartz, I. & Harry, B. (1998) Participatory research: New approaches to the research to practice dilemma. *Journal of the Association for Persons with Severe Handicaps*, 23 (3), 165–177. Available at https://doi.org/10.2511/rpsd.23.3.165.

Minkler, M. & Wallerstein, N. (eds.) (2011) *Community-based participatory research for health: From process to outcomes*. 2nd ed. Jossey-Bass.

Miyashita, H. (2024) Using online discussion forums in blended learning design to advance higher order thinking. *Asian Journal of Distance*

Education, 19 (1), 82-98. Available at
https://asianjde.com/ojs/index.php/AsianJDE/article/view/738.

Moran, V.H., Maqsood, M., Fatima, S., Ullah, M., Cruz-Rodriguez, M., & Zaman, M. (2023) Participatory action research to co-design a culturally appropriate COVID-19 risk communication and community engagement strategy in rural Pakistan. *Frontiers in Public Health*, 11, Article 1160964. Available at https://doi.org/10.3389/fpubh.2023.1160964.

Mori, A. (2019) Facilitating collaboration between Japanese high schools and universities: a qualitative exploration of the role of education outreach coordinators. *Frontiers in Education*, 4, 1393183. Available at https://doi.org/10.3389/feduc.2024.1393183.

Park, P. (1999) People, knowledge, and change in participatory research. *Management Learning*, 30 (2), 141–157. Available at https://doi.org/10.1177/1350507699302003.

Pedraza Borbon, A. (2024) Supporting English language learners through peer collaboration and scaffolding. *Journal of English Learner Education*, 16 (2), Article 4. Available at https://stars.library.ucf.edu/jele/vol16/iss2/4.

Pitura, J. (2024) Participatory action research: advocacy and activism for promoting social justice in and through CALL. *Language Learning & Technology*, 28 (1). Available at https://doi.org/10.1080/09588221.2024.2310290.

Rampton, M. B. H. (1990) Displacing the "native speaker": Expertise, affiliation, and inheritance. *ELT Journal*, 44 (2), 97–101. Available at https://doi.org/10.1093/eltj/44.2.97.

Reason, R. (1998) Three approaches to participative inquiry. In Denzin, N.K. & Lincoln, Y.S. (Eds.) *Strategies of qualitative research*, pp. 261–291. Sage.

Reason, P. (2006) Choice and quality in action research practice. *Journal of Management Inquiry*, 15 (2), 187–203. Available at https://doi.org/10.1177/1056492606288074.

Reason, P. & Bradbury, H. (Eds.) (2013) *The SAGE handbook of action research: Participative inquiry and practice* (2nd ed.). SAGE Publications.

Reeder, K., Macfadyen, L. P., Roche, J. & Chase, M. (2004) Negotiating cultures in cyberspace: Participation patterns and problematics. *Language Learning and Technology*, 8 (2), 88–105. Available at https://core.ac.uk/reader/84320980.

Revans, R.W. (1982) *The origins and growth of action learning*. Chartwell-Bratt.

Richardson, E. V., Nagata, S., Hall, C., Akimoto, S., Barber, L., & Sawae, Y. (2023) Developing a socially-just research agenda for inclusive physical education in Japan. *Quest*, 75 (4), 361–378. Available at https://doi.org/10.1080/00336297.2023.2206578.

Ridings, C., Gefen, D. & Arinze, B. (2006) Psychological barriers: Lurker and poster motivations and behavior in online communities. *Communications of the Association for Information Systems*, 18 (1), 329–354. Available at https://doi.org/10.17705/1CAIS.01816.

Robbins, M. & Fairbank, J. (2023) Asynchronous online discussion forums: Effective undergraduate and graduate course approaches. *Journal of Online Learning*, 15 (2) 112-125. Available at https://files.eric.ed.gov/fulltext/EJ1407501.pdf.

Rowe, G. & Wright, G. (1999) The Delphi technique as a forecasting tool: Issues and analysis. *International Journal of Forecasting*, 15 (4), 353–375. Available at https://doi.org/10.1016/S0169-2070(99)00018-7.

Sandelowski, M. (2002) Reembodying qualitative inquiry. *Qualitative Health Research*, 12 (1), 104–115. Available at https://doi.org/10.1177/1049732302012001008.

Sattler, C., Rommel, J., Chen, C., García-Llorente, M., Gutiérrez-Briceño, I., Prager, K., Reyes, M.F., Schröter, B., Schulze, C., van Bussel, L.G. & Loft, L, (2022) Participatory research in times of COVID-19 and beyond: Adjusting your methodological toolkits. *One Earth*, 5 (1), 62-73. Available at https://doi.org/10.1016/j.oneear.2021.12.006.

Skulmoski, G. J., Hartman, F. T. & Krahn, J. (2007) The Delphi Method for graduate research. *Journal of Information Technology Education*, 6 (1), 1–21. Available at https://doi.org/10.28945/199.

Stoecker, R. (1999) Are academics irrelevant? Roles for scholars in participatory research. *American Behavioral Scientist*, 42 (5), 834–848. Available at https://doi.org/10.1177/00027649921954561.

Stringer, E. T. (1996) *Action research: A handbook for practitioners*. Sage.

Stringer, E. T. (2014) *Action Research*. 4th ed. Sage.

Tawalbeh, M. & Al-Husban, N. (2023) EFL students' perspectives on activities designed for asynchronous discussion forums: Transformative practices. *International Journal of Technology in Education*, 6 (3), 507–520. Available at https://doi.org/10.46328/ijte.519.

Tsubono, K. & Mitoku, S. (2023) Public school teachers' occupational stress across different school types: a nationwide survey during the prolonged COVID-19 pandemic in Japan. *Frontiers in Public Health*, 11, Article 1287893. Available at https://doi.org/10.3389/fpubh.2023.1287893.

Upreti, Y.R., Devkota, B. & Maharjan, S.K. (2024). Participatory action research: An emergent research methodology in health education and promotion. *Journal of Health Promotion*, 12 (1), 1-8. Available at https://doi.org/10.3126/jhp.v12i1.72690.

Vallenga, D., Grypdonck, M. H. F., Hoogwerf, L. J. R. & Tan, F. I. A. (2009) Action research: What, why, and how? *Acta Neurologica Belgica*, 109 (2), 81–90. Available at https://pubmed.ncbi.nlm.nih.gov/19681439/.

Voltage Control (2024) *Transforming education: Effective change management in schools*. Available at https://voltagecontrol.com/articles/transforming-education-effective-change-management-in-schools/.

Wang, M. & Zhang, J. (2023) Understanding teachers' online professional learning: A community of inquiry perspective on the role of Chinese middle school teachers' sense of self-efficacy, and online learning achievement. *Heliyon*, 9 (6), e16932. Available at https://doi.org/10.1016/j.heliyon.2023.e16932.

Wright, S. (2024) *Emancipatory participation for online engagement*. Routledge.

Xie, Y. (2023) The impact of online office on social anxiety among primary and secondary school teachers. *Frontiers in Psychology*, 14, Article 1154460. Available at https://doi.org/10.3389/fpsyg.2023.1154460.

Xie, J., Adjei, M. & Correia, A.P. (2024) The impact of different instructor participation approaches in asynchronous online discussions on student performance. *Online Learning*, 28 (4), 34–56. Available at https://files.eric.ed.gov/fulltext/EJ1455357.pdf.

Yeow, A., Johnson, S. & Faraj, S. (2006) Lurking: Legitimate or illegitimate peripheral participation. In *Proceedings of the Twenty-Seventh International Conference on Information Systems*, pp. 967–982. Milwaukee, Wisconsin. Available at https://aisel.aisnet.org/cgi/viewcontent.cgi?article=1183&context=icis2006.

Yildiz, S. & Bichelmeyer, B. A. (2003) Exploring electronic forum participation and interaction by EFL speakers in two web-based graduate-level courses. *Distance Education*, 24 (2), 176–193. Available at https://doi.org/10.1080/0158791032000127464.

Zhang, M. & Liu, Q. (2023) Synchronous and asynchronous online collaborative writing: Effects of task modality on utility and engagement in Chinese as a foreign language learners. *Foreign Language Annals*, 56 (4), 755–774. Available at https://doi.org/10.1111/flan.12704.

Zhang, W. & Hu, W. (2025) Effects of perceived English teacher support on student engagement among Chinese EFL undergraduates: L2 motivational self-system as the mediator. *Frontiers in Psychology*, 12, 769145. Available at https://www.frontiersin.org/journals/psychology/articles/10.3389/fpsyg.2025.1607414/full.

Zhao, Y., Li, X. & Kang, H. (2025) Linking distributed leadership to teachers' innovation: Chain mediating roles of commitment and collaboration in Chinese schools. *PLoS ONE*, 20 (9) e0333118. Available at https://doi.org/10.1371/journal.pone.0333118.

Zhong, J. Ismail, L. & Lin, Y. (2025) Investigating EFL students' engagement in project-based speaking activities: from a multi-dimensional perspective. *Frontiers in Psychology*, 16. Available at https://www.frontiersin.org/journals/psychology/articles/10.3389/fpsyg.2025.1598513/full.

Biography

Stuart D. Warrington, Ed.D., is a Professor and Head of the Self-Access Center (SAC) Committee in the Department of British and American Studies at Nagoya University of Commerce and Business (NUCB), Japan. He has extensive experience in language education, teacher and researcher development, with a particular focus on self-access language learning and the autonomy and agency of learners, learning advisors, and managers. His research explores the evolving professionalism and ongoing professional development of self-access learning managers and advisors, the pedagogical design and practical use of online and physical learning spaces, and ecological approaches to learner autonomy and agency. He has published in international journals, delivered conference, workshop and symposium presentations, and actively contributes to professional organizations supporting self-access language learning worldwide.

ORCID: 0000-0001-5140-808X

Book review - The politics of action research: A story telling inquiry

Yedida Bessemer

Hill, G., & Rixon, A. (2024). *The Politics of Action Research: A Storytelling Inquiry*. Cambridge Scholars Publishing.

In *The Politics of Action Research: A Storytelling Inquiry* (2024), Geof Hill and Andrew Rixon conducted a comprehensive analysis of action research through the lens of power, voice, and political dynamics. The authors position their work as both a methodological contribution and a critical intervention. They claimed that action research is inherently political and is embedded in relations of power that shape who can research, what counts as knowledge, and whose voices are legitimized in academic and organizational spaces. By drawing on storytelling as an inquiry methodology, Hill and Rixon collected and analyzed narratives from action research practitioners across multiple continents and contexts. The authors illuminated themes of silencing, empowerment, gatekeeping, and meaning-making. The book's central assertion is that understanding the political dimensions of action research is essential for practitioners, supervisors, and institutional leaders who seek to direct and transform the complex terrain of participatory inquiry.

The book affirms a classic, cyclical view of action research: plan, act, observe, reflect, but reframes these cycles as politically situated rather than neutral procedural steps. The authors' use of Bob Dick's foreword effectively sets the stage, distinguishing between viewing action research as "apolitical" versus recognizing it as embodying a different *kind* of politics; in other words, one characterized by egalitarian participation and collaborative negotiation rather than hierarchical control. This framing resonates throughout the book's thematic chapters, which explore how practitioners experience empowerment and disempowerment, navigate gatekeeping structures (such as ethics committees and journal review processes), and develop resilience through creative methodological adaptations.

The book's practical contribution to the field lies in three areas: methodological guidance, navigational strategies, and legitimation of practitioner knowledge. The first contribution is that Hill and Rixon model storytelling as inquiry throughout the text, making their analytical process transparent. Chapter 6 explicates how stories are collected, authenticated, and analyzed through thematic coding and "six degrees of separation." This approach traces connections across narratives to identify patterns and tensions. This methodological transparency is particularly valuable for novice action researchers seeking to understand how narrative data can generate rigorous, theoretically grounded insights.

The second contribution is the book's offer of navigational strategies for practitioners working "in the swamp," as the readers can find in Chapter 10, Schön's (1983) metaphor for messy, complex practice contexts. Also, stories of researchers challenging third-person writing conventions (Somekh, 1995), developing alternate quality criteria (Winter, 2002), and incorporating artful inquiry methods (Lloyd, Gordillo) provide models for creative resistance to constraining academic norms. These examples demonstrate that methodological innovation is not simply technical but political, and it is a claim to epistemic legitimacy for diverse ways of knowing.

The third contribution is that the book legitimizes practitioner expertise by centering lived experience. The use of first-person narratives and visual artifacts (e.g., Paul Costelloe's provenance collage, Ekaterina Timokhina's wing imagery) honors multiple modes of knowledge expression. It embodies action research's commitment to "extended epistemologies" (Heron & Reason, 2008). This representational choice counters academic gatekeeping that privileges forms of scholarly discourse.

With that in mind, the book's practical guidance would benefit from more explicit synthesis. Even though the authors provided multiple examples, readers seeking structured protocols for addressing gatekeeping, designing ethics applications for participatory research, or building institutional support for action research must extract insights from dispersed narratives. A concluding chapter offering consolidated recommendations for practitioners, supervisors, and institutional leaders would have been an additional benefit to readers.

The Politics of Action Research demonstrates strengths in its scope, authenticity, and reflexivity. The global reach of contributors reflects the international evolution of action research. The authors' commitment to participant authentication and their explicit discussion of anonymity honor ethical principles of collaborative research. In addition, the book's theoretical framing is sophisticated, as it interlaces critical theory (Habermas, Gramsci), practice theory (Schön, Wittgenstein), and creativity studies (Gardner, Haseman) to explain how power operates in research contexts.

However, as with most books, there are limitations that warrant the readers' attention. While the book critiques gatekeeping, its own theoretical density and academic style may ironically limit accessibility for some practitioner audiences, especially for those new to action research or working outside university settings. The extensive literature reviews and philosophical discussions, while scholarly and rigorous, could be balanced with more synthesized "takeaways."

The thematic organization sometimes creates redundancy. Concepts like empowerment, voice, and resistance appear across multiple chapters (7, 8, 10, 11), and tighter editorial integration could strengthen coherence. For example, the distinction between "silencing" and "gatekeeping" as separate themes found in chapters 7 and 9 is not always clear, as both address institutional barriers to legitimacy.

While the book acknowledges diversity in action research traditions, it gives limited attention to decolonial and Indigenous methodologies. Although Fals-Borda's critique of Eurocentrism is cited, deeper engagement with non-Western epistemologies and their relationship to action research politics would strengthen the analysis.

Finally, the future-oriented discussion in Chapter 12 feels somewhat brief given the book's length. While Hilary Bradbury's concept of Action-Oriented Research for Transformations (ART) and the discussion of "clusters" for sustaining action research communities are promising, more concrete guidance for building institutional infrastructure and responding to contemporary challenges (e.g., artificial intelligence, climate crisis, digital transformation) would enhance relevance.

Despite these limitations, multiple audiences will benefit from this book. The first group, doctoral candidates undertaking action research dissertations, will find validation and guidance in the stories of struggle and persistence. The detailed accounts of ethics approval challenges, examination processes, and publication rejections prepare students for institutional hurdles while modeling resilience and creative problem-solving.

The second group, supervisors and program directors, will benefit from understanding the political dynamics their students navigate, which could inform more supportive supervisory practices and program design. The book implicitly argues for institutional change, urges universities to develop action-research-literate ethics committees, and recognizes diverse forms of scholarly contribution while creating supportive structures for participatory inquiry.

The third group, organizational leaders and community-based researchers, will appreciate the book's demonstration that action research offers powerful tools for democratic participation and transformational change, even as it requires courage to challenge established norms. Examples from business schools, healthcare, education, and humanitarian contexts illustrate the versatility of action research.

The book's focus is academia and doctoral education and academic publishing, centered more on institutional concerns than grassroots applications, and its consequential impact will reflect this focus.

Hill and Rixon's (2024) work suggests three productive guidelines for action research scholarship, such as, Institutional ethnographies of action research practice, a systematic study of how universities, funding bodies, and professional associations shape (or constrain) action research through policies, review processes, and resource allocation, which would extend this book's insights into organizational structures supporting or hindering participatory inquiry. Moreover, an intersectional analysis of power in action research could examine how race, gender, class, Global North/South positionality, and other identity dimensions intersect with epistemic power. How do women of color, Indigenous researchers, or scholars from the Global South experience action research gatekeeping differently than the predominantly white, Western practitioners featured here? Furthermore, action research pedagogy and capacity building, which involves a systematic investigation of how action research is taught, mentored, and sustained, particularly in institutions without established traditions, would support wider adoption.

All in all, Hill and Rixon's (2024) *The Politics of Action Research* expands the discourse beyond method into meaning, identity, and power. Its storytelling methodology reveals action research not as a neutral technique but as a political practice rooted in collaboration, ethical commitment, and reflective transformation. In doing so, it contributes substantively to the theory, practice, and pedagogy of action research, offering a generative platform for scholars and practitioners engaged in real-world improvement.

Adjunct Professor Yedida Bessemer
University of Charleston, West Virginia, USA

Membership information and article submissions

Membership categories

Membership of Action Learning, Action Research Association Ltd (ALARA) takes two forms: individual and organisational.

ALARA individual membership

Members of the ALARA obtain access to all issues of the *Action Learning and Action Research Journal* (*ALARj*) twelve months before it becomes available to the public.

ALARA members receive regular emailed Action Learning and Action Research updates and access to web-based networks, discounts on conference/seminar registrations, and an on-line membership directory. The directory has details of members with information about interests as well as the ability to contact them.

ALARA organisational membership

ALARA is keen to make connections between people and activities in all strands, streams and variants associated with our paradigm. Areas include Action Learning, Action Research, process management, collaborative inquiry facilitation, systems thinking, Indigenous research and organisational learning and development. ALARA may appeal to people working at all levels in any kind of organisational, community, workplace or other practice setting.

ALARA invites organisational memberships with university schools, public sector units, corporate and Medium to Small Business, and community organisations. Such memberships include Affiliates. Details are on our membership link on our website (https://alarassociation.org/membership/Affiliates).

Become a member of ALARA

An individual Membership Application Form is on the last page of this Journal or individuals can join by clicking on the Membership Application button on ALARA's website. Organisations can apply by using the organisational membership application form on ALARA's website.

For more information on ALARA activities and to join
Please visit our web page:
https://www.alarassociation.org/user/register
or email admin@alarassociation.org

Journal submissions criteria and review process

The *ALARj* contains substantial articles, project reports, information about activities, creative works from the Action Learning and Action Research field, reflections on seminars and conferences, short articles related to the theory and practice of Action Learning and Action Research, and reviews of recent publications. *ALARj* also advertises practitioners' services for a fee.

The *ALARj* aims to be of the highest standard of writing from the field in order to extend the boundaries of theorisation of the practice, as well as the boundaries of its application.

ALARA aims *ALARj* to be accessible for readers and contributors while not compromising the need for sophistication that complex situations require. We encourage experienced practitioners and scholars to contribute, while being willing to publish new practitioners as a way of developing the field, and introduce novice practitioners presenting creative and insightful work

We will only receive articles that have been proof read, comply with the submission guidelines as identified on *ALARj*'s website, and that meet the criteria that the reviewers use. We are unlikely to publish an article that describes a project simply because its methodology is drawn from our field.

ALARA intends *AlARj* to provide high quality works for practitioners and funding bodies to use in the commissioning of works, and the progression of and inclusion of action research and action learning concepts and practices in policy and operations.

ALARj has a substantial international panel of experienced Action Learning and Action Research scholars and practitioners who offer double blind and transparent reviews at the request of the author.

Making your submission and developing your paper

Please send all contributions in Microsoft Word format to the Open Journal Systems (OJS) access portal: https://alarj.alarassociation.org.

You must register as an author to upload your document and work through the electronic pages of requirements to make your submission. ALARA's Managing Editor or Issue Editor will contact you and you can track progress of your paper on the OJS page.

If you have any difficulties or inquiries about submission or any other matters to do with ALARA publications contact the Managing Editor on editor@alarassociation.org.

For the full details of submitting to the *ALAR Journal*, please see the submission guidelines on ALARA's web site https://alarassociation.org/publications/submission-guidelines/alarj-submission-guidelines.

Guidelines

ALARj is devoted to the communication of the theory and practice of Action Learning, Action Research and related methodologies generally. As with all ALARA activities, all streams of work across all disciplines are welcome. These areas include Action Learning, Action Research, Participatory Action Research, systems thinking, inquiry process-facilitation, process management, and all the associated post-modern epistemologies and methods such as rural self-appraisal, auto-ethnography, appreciative inquiry, most significant change, open space technology, etc.

In reviewing submitted papers, our reviewers use the following criteria, which are important for authors to consider:

Criterion 1: How well are the paper and its focus both aimed at and/or grounded in the world of practice?

Criterion 2: How well are the paper and/or its subject explicitly and actively participative: research with, for and by people rather than on people?

Criterion 3: How well do the paper and/or its subject draw on a wide range of ways of knowing (including intuitive, experiential, presentational as well as conceptual) and link these appropriately to form theory of and in practices (praxis)?

Criterion 4: How well does the paper address questions that are of significance to the flourishing of human community and the more-than-human world as related to the foreseeable future?

Criterion 5: How well does the paper consider the ethics of research practice for this and multiple generations?

Criterion 6: How well does the paper and/or its subject aim to leave some lasting capacity amongst those involved, encompassing first, second and third person perspectives?

Criterion 7: How well do the paper and its subject offer critical insights into and critical reflections on the research and inquiry process?

Criteria 8: How well does the paper openly acknowledge there are culturally distinctive approaches to Action Research and Action Learning and seek to make explicit their own assumptions about non-Western/ Indigenous and Western approaches to Action Research and Action Learning

Criteria 9: How well does the paper engage the context of research with systemic thinking and practices

Criterion 10: How well do the paper and/or its subject progress AR and AL in the field (research, community, business, education or otherwise)?

Criterion 11: How well is the paper written?

Article preparation

ALARj submissions must be original and unpublished work suitable for an international audience and not under review by any other publisher or journal. No payment is associated with submissions. Copyright of published works remains with the author(s) shared with Action Learning, Action Research Association Ltd

While *ALARj* promotes established practice and related discourse *ALARj* also encourages unconventional approaches to reflecting on practice including poetry, artworks and other forms of creative expression that can in some instances progress the field more appropriately than academic forms of writing.

Submissions are uploaded to our Open Journal System (OJS) editing and publication site.

The reviewers use the OJS system to send authors feedback within a 2-3 month period. You will receive emails at each stage of the process with feedback, and if needed, instructions included in the email about how to make revisions and resubmit.

Access to the journal

The journal is published electronically on the OJS website.

EBSCO and InformIT also publish the journal commercially for worldwide access, and pdf or printed versions are available from various online booksellers or email admin@alarassociation.org.

For further information about the *ALAR Journal* and other ALARA publications, please see ALARA's web site http://www.alarassociation.org/publications.

Individual Membership Application Form

This form is for the use of individuals wishing to join ALARA.
Please complete <u>all</u> fields.

Name
Title	Given Name		Family Name

Residential Address
Street	Town / City	Postcode / Zip
Country		

Postal Address
Street	Town / City	Postcode / Zip
State	Country	

Telephone
Country Code	Telephone Number

Mobile Telephone
Country Code	Mobile Number

Email
Email Address

Experience (Please tick most relevant)
- [ ] No experience yet
- [ ] 1 – 5 years' experience
- [ ] More than 5 years' experience

Interests (Please tick all relevant)
- [ ] Education
- [ ] Health
- [ ] Community / Social Justice
- [ ] Indigenous Issues
- [ ] Gender Issues
- [ ] Organizational Development

Are you eligible for concessional membership?
If you are a full-time student, retired or an individual earning less than AUD 20,000 per year; about USD 13,750 (please check current conversion rates), you can apply for concessional membership.

Do you belong to an organization that is an Organizational Member of ALARA?
If you are a member of such an organization, you can apply for the Reduced Membership Fee. Please state the name of the Organizational Member of ALARA in the box below.

Payment
We offer a range of payment options. Details are provided on the Tax Invoice that we will send to you on receipt of your membership application.

[AusPost billpay] [Bpay] [MasterCard] [VISA]

If you want to join and pay online, please go to https://www.alarassociation.org and click on the Membership Application button (lower right). Alternatively, please complete and return this form to us.

By Post
ALARA Membership
PO Box 182 Greenslopes
Queensland 4120
AUSTRALIA

By FAX
+ 61 (7) 3342 1669

By Email
admin@alarassociation.org

Annual Membership Fees (Please select one)

Full Membership		Concessional Membership	
[ ] AUD 143.00	Developed Country	AUD 71.50	
[ ] AUD 99.00	Emerging Country	AUD 49.50	
[ ] AUD 55.00	Developing Country	AUD 27.50	

Reduced Membership Fee, as I belong to an Organizational Member of ALARA	Developed	AUD 71.50
	Emerging	AUD 49.50
	Developing	AUD 27.50

Organization's name

Privacy Policy
By submitting this membership form, I acknowledge that I have read, understood and accept ALARA's Privacy Policy https://www.alarassociation.org/sites/default/files/docs/policies/ALARA_PrivacyPolicy11_1.pdf.

ALARA will acknowledge receipt of your application and send you an invoice or receipt of payment. You will receive an email confirming activation of your account, and details on how you can access website functions.

ALARA is a global network of programs, institutions, professionals, and people interested in using action learning and action research to generate collaborative learning, training, research and action to transform workplaces, schools, colleges, universities, communities, voluntary organisations, governments and businesses.

ALARA's vision is to create a more equitable, just, joyful, productive, peaceful and sustainable society by promoting local and global change through the wide use of Action Learning and Action Research by individuals, groups and organisations.

www.ingramcontent.com/pod-product-compliance
Ingram Content Group UK Ltd.
Pitfield, Milton Keynes, MK11 3LW, UK
UKHW020719050526
12271UKWH00018B/218